Also by Daniel Stih

Healthy Living Spaces
Top 10 Hazards Affecting Your Health

Mold Money
*How to Save Thousands of Dollars on Mold
Remediation and Make Sure the Mold is Gone*

How to Build a Healthy Home
*And Prevent the Negative Impacts on Your Health
That Can Result from Poorly Executed Green
Building Initiatives*

Unplugged

How to Find and Get Rid of

EMFs in Your Home

Inquiries regarding requests to reprint
parts should be addressed to the pub-
lisher at the following address:

Healthy Living Spaces®
369 Montezuma Ave #169
Santa Fe, NM 87501

Dedicated to Helmut Ziehe

WARNING!

When in doubt—STOP! Call an electrician.

Enough voltage can burn your skin, shock you, or stop your heart.

The author does not assume responsibility for actions taken based on reading this book. Readers should hire an electrician to help them identify and repair electrical issues causing high magnetic or electric fields.

TABLE OF CONTENTS

Introduction

Electromagnetic Fields Demystified

What EMFs Can Do to You

Meters

How to Take Readings

What the Readings Mean

Determining Why a Field is High

Getting Rid of Problem Fields

Shielding

Supplementary Information

Introduction

Electricity is a fact of life. It may be difficult to get away from. This book contains practical advice for reducing and minimizing exposure. A few small changes may make a world of difference. The term electromagnetic fields (EMFs) is technical. I've tried to keep it simple. If you find yourself overwhelmed, try skipping to another chapter. The next topic may be easier to understand. Although it seems incredible that an EMF could make someone irritable, uncomfortable, or unwell, I have seen differences in how clients feel after reducing or eliminating EMFs. If you do not have a meter, you can still try the recommendations in this book. My apologies to electrical engineers who find my explanations on physics and electricity over-simplified. This book is to help the average person understand EMFs and, more importantly, how to get rid of them.

Electromagnetic Fields Demystified

What They Are and Where They Come From

EMFs Demystified

Voltage

Voltage (V) is a measure of how much energy electrons are charged with. The amount of voltage determines if the electrons go (flow) and make it around the track. The track is the wire; it might be a long wire, or the track might be uphill. On a battery, the starting line is the positive terminal. The finish line is the negative terminal, back at the battery, after completing the loop. Electrons returning to the battery repeat the circuit (go around the track again) until the battery runs out of energy to give them.

In a house, the amount of voltage required to make things work depends on the device. A motor in a washing machine for example, uses all the voltage. Electronic devices require less power and use transformers to lower the voltage.

Why 120 V was chosen is not clear. When power lines were being developed, Edison's General Electric Company worked on a system of 110 V. Tesla devised a system of 240 V, which is more efficient than 110 V. With the backing of the Westinghouse, Tesla's system became the standard. Tesla is said to have compromised, and he reduced the voltage to 120 V for safety and political reasons.

More than half of the world uses electricity at 240 V. It's too expensive for the United States to change. Europe switched to 230 V at the end of World War II and to 50 Hertz (Hz), instead of the American 60 Hz, to fit the metric system. Some European countries use 220V.

Voltage

EMFs Demystified

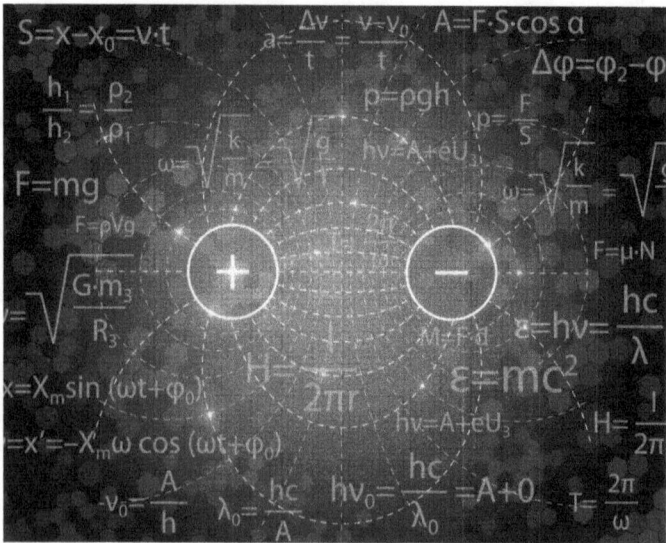

Electric Fields

An electric field is the intensity associated with being close to something with a high voltage. While the scientific units are volts per meter (V/m), it might be easier to understand as volts per foot.

The strength of a field decreases with the square of the distance. Thus, if you walk two feet away from a source, the field will be 1/4, or 25%, lower. Walk three feet away and the field will be 1/9, or almost 90%, lower. Consider this rule when trying to keep a safe distance from sources that cause elevated electric fields, such as power lines and house wiring. In your house, wires hidden in the walls create electric fields, even when the lights are off. An electric field can only be eliminated by shutting off the power. As it is not practical to shut power off, I will provide suggestions for lowering the fields in later chapters.

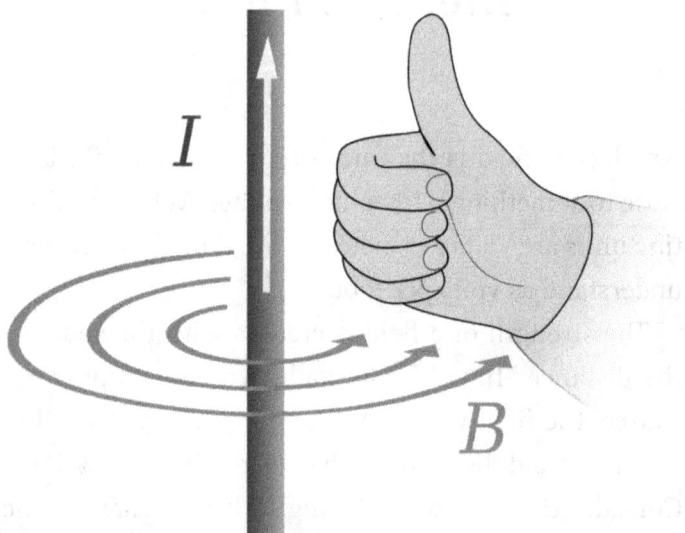

When current (I) flows on a wire, a magnetic field (the B circle) is created around it. A good question is why "I" and "B" were chosen as the letters. There are other variables in the equations that govern electric and magnetic fields. Each variable is represented by a letter. When Maxwell formulated the equations, he started with the term "magnetic potential," giving it the letter A. The next term he assigned a letter to was "magnetic field," to which he gave the letter B. By the time he got to "current," he was at the letter I.

Magnetic Fields

Once a light or something is turned on, a current flows, creating a magnetic field. <u>Magnetic fields are only created when current flows.</u> When current is not flowing, there is no magnetic field. No one knows why. It's a phenomenon of nature, one that is used to make electricity. If you pass a magnet around a loop of wire, current flows.

If the wires in a house are balanced, the magnetic fields created by the current flowing on them is canceled. Current going to where it is needed creates a magnetic field. That field is canceled by current flowing back, in the opposite direction, on the same set of wires. Elevated magnetic fields occur when current that goes to where it's needed in a home does not come back on the same set of wires. The National Electric Code was written to avoid mistakes and to keep current "balanced." But wiring a house is complex and it's a wonder that more mistakes are not made.

Computer simulation of Earth's magnetic field created by NASA. The field is created by an electric current in molten iron located in the outer core. The field extends into space where it interacts with solar wind, streams of charged particles emanating from the sun.

The Earth's Natural Magnetic Field

Earth has a magnetic field powerful enough to move the needle in a compass. Earth's field is between 0.3 and 0.6 Gauss (G), larger than magnetic fields created by electric power. Fields near power lines are measured in milli-Gauss (mG), 1,000 times lower.

How is it possible that fields from electric power affect the human body if Earth's natural field is higher than that from power lines? One explanation may be that Earth's field is the result of current that flows in a constant direction. Electricity delivered by the electric company changes directions 60 times each second. This is known as alternating current (AC), and is done to save energy. It is possible that the constant changing of the direction of the field has an effect on health. Another explanation may be the interactions between man-made and natural fields. Researchers suspect a resonance effect. Resonance occurs when an oscillation is reinforced by something else vibrating at the same rate.

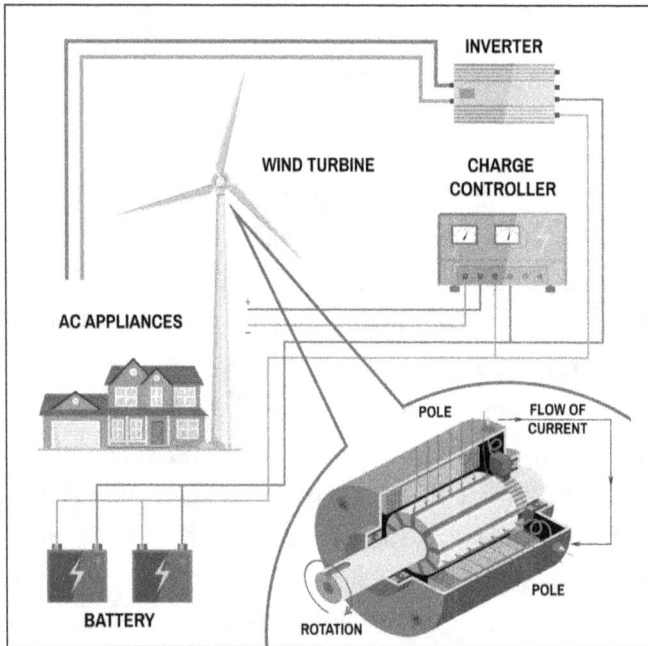

INVERTER

WIND TURBINE

CHARGE
CONTROLLER

AC APPLIANCES

POLE

FLOW OF
CURRENT

POLE

BATTERY

ROTATION

Inside the generator of a windmill, magnets spin around copper wires. As if by magic, an electric current is made to flow on the wires with each pass of the magnets.

Electric Power

If you take a loop of wire and pass a magnet through it, current will flow. It's a physical phenomenon. The amount of current depends on the size and number of the magnets and copper loops. To generate electric power, many magnets are repeatedly passed through big loops of copper as fast as possible.

It's more energy efficient if, before delivering the power, the electric company increases the voltage. This is because energy is lost to heat from the resistance of the wires. Think of the wires that generate heat in a toaster oven. By increasing the voltage, not as much current needs to flow to get the same amount of energy to homes. A power distribution line could be at 69,000 V. The power in a home is at 120 V. Transformers (the round tubs on utility poles or the green boxes in front yards) lower the voltage as it reaches homes.

It would be healthier if the direction the current flowed did not alternate back and forth. In the 1800s, the technology did not exist to convert current to high voltages without using a type of transformer that alternated

Transformers on electrical poles lower the voltage of electricity as it's delivered to homes.

the direction of the current. With the advent of semi-conductors in the 1970s, it became cheaper to build High Voltage Direct Current (HVDC) power distribution systems. These are more expensive. Parts of Europe have started to implement HVDC systems to connect countries. HVDC systems experience less loss of energy over long distances than systems that alternate the direction of the current.

Phones, computers, and other electronic devices have circuits in their chargers which convert AC to DC. The semiconductor in the circuit is called a rectifier.

With alternating current, if you could slow time, you would see the lights in a house pulse like a heartbeat. For a moment, the lights will be bright. As the current changes direction, the voltage will drop to zero, and the lights will dim. It will be dark for a moment before the current increases in the opposite direction and the lights glow bright again. This repeats 60 times a second.

For current to flow, the current needs somewhere to go. Appliances, light bulbs, and computers—these take energy from electrons; they don't keep the electrons. To complete the circuit, electrons flow back to the fuse box and then to the power company.

If you cut open a wire, you will see three smaller wires. The black is called the hot wire. It carries high voltage electricity. The white is called the neutral wire. The neutral has only one-half of a volt. The rest is used for power. The neutral is what electrons are supposed to travel back on, paired with the hot wire they arrived on. The green, or bare wire, is called the ground wire. Current is not supposed to be on the ground wire.

The purpose of the ground is to have a reference to ensure that every house is at the same 120 V. Voltage is like speed. If I tell you I am going 400 miles per hour, you might not understand. Flying in a jet, you can go 400 mph relative to the ground. If you get out of your seat and walk to the lavatory, it's not that fast relative to the floor in the airplane. Things that use electricity are designed to operate at 120 V relative to the ground (Earth). The ground is considered to be at zero volts. The ground wire in a house is connected to metal water pipes and a three-foot copper stake driven into or buried in the ground next to the fuse box.

Some electricians believe the reason for connecting the ground wire to metal pipes is to prevent shock. Grounding does not prevent shock. If it did, there would be no need to install ground fault interrupters (GFIs) in kitchens, bathrooms, and garages. A GFI monitors the current and shuts off the circuit if any current fails to return on the circuit.

Electric Power

Mounting of a steam turbine produced by Siemens, Germany.

Author: Siemens Pressebild.
https://new.siemens.com/global/en.html

Most electric power is produced by steam-electric power plants.
Water is heated, turns into steam, and spins a turbine which drives
an electrical generator. Coal, nuclear, and many natural gas power
plants are steam-electric. Non-steam electric plants include
hydroelectric, gas turbine plants, photovoltaic, and wind turbines.

Ground Currents

As a result of the way things are grounded, a portion of the electricity delivered to homes flows back to the electric company though the ground. Large grounding grids are buried under power substations. The conductivity of the materials in the earth determines the magnitude of the current. Saturated soil, lakes, and streams may carry more current than dry soil. This may have health ramifications for all living things.

Another way Earth is used by the electric company is for disposal. Large power plants are unable to change their outputs quickly with demand. Sometimes, it's necessary to shunt electrical current into the earth until the output can be adjusted to meet the demand.

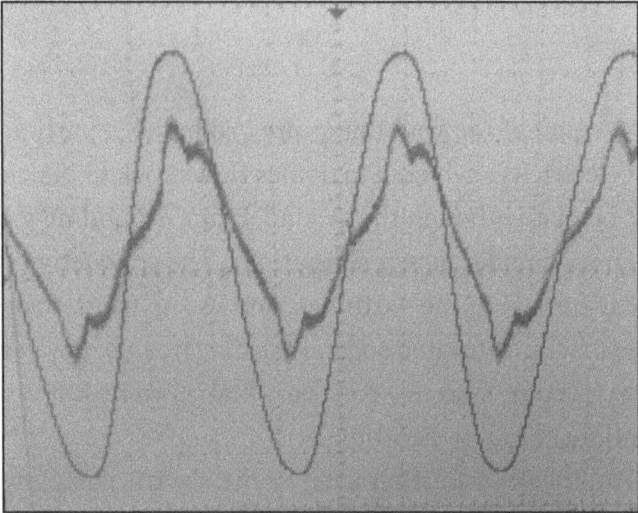

Voltage on a wire in a home displayed on an oscilloscope.
The big, smooth wave is the hot wire at an average of 120 V,
going up and down from +170 to -170 V, as the current
alternates directions. The thicker, dark wave is the voltage
(1/2 volt) on the neutral wire, the wire used to return the
current to the electric company. The wave has been made
choppy by electronic devices.

Dirty Electricity

Dirty Electricity is a term used to describe electricity not in perfect shape. As an electric company delivers power, the wave goes up and down. When the quality of the power is clean, the wave is smooth and rounded. When it is interfered with, it becomes choppy.

Choppy waves are caused by electronics—devices that can't handle 120 V. Devices that need less power, such as cell phones, computers, digital clocks, and LED displays, use transformers to lower the voltage. Think of it as if they are taking a bite out of the pie. Some want 1/3; others want 1/6.

EMFs Demystified

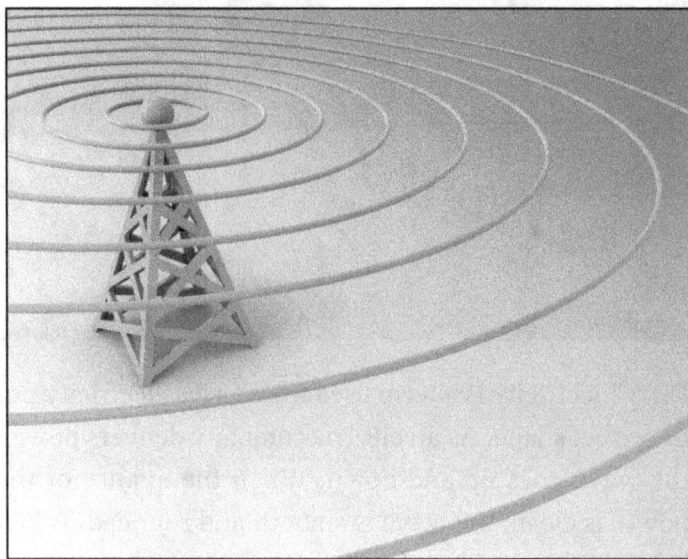

Radio Frequencies

Radio frequencies (RFs) refer to frequencies higher than those created by electric power. Sources include cell phones, WiFi, radar, radio, and TV. The term electromagnetic field is literal. High frequency fields are alternating electric and magnetic fields that travel through space. An electric field is initiated at the transmitter, the antenna. The electric field creates a magnetic field in space, which creates an electric field in space, which creates a magnetic field. This goes back and forth, propagating in space. As the wave reaches its destination, the receiving antenna detects the electric field in space. The frequency of the back and forth is the frequency of the station. For example, 108.5 MHz (FM radio) is a wave with crests and troughs, cycling up and down 108.5 million times per second.

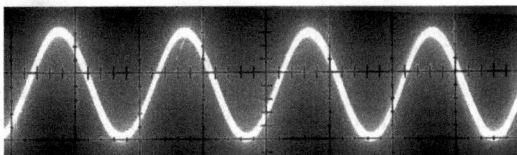

UNITED

STATES

FREQUENCY

ALLOCATIONS

THE RADIO SPECTRUM

See any blank space? The spectrum is full. The Federal Communications Commission (FCC) has allocated frequencies for TV, radio, cell phones, Bluetooth®, WiFi,radar, fire, police, weather, airport controllers, satellites, ham radios, government, and the military.

Radio Frequencies

300 kHz

3 MHz

30 MHz

300 MHz

3 GHz

30 GHz

300 GHz

History of high frequency EMF waves

James Clerk Maxwell (1831-1879) formulated "Maxwell's Equations," the mathematics for how electromagnetic waves propagate through space. He showed that light is the same phenomenon at different wavelength frequencies. Tesla is said to have considered the equations useless, since light, he said, could not transmit further than line of sight. Tesla proposed a system for transmitting wireless power through the earth. Some believe Tesla invented a technology that would have provided free energy. It's difficult to separate fact from fiction. Upon his death, the government seized his files, and some are still missing.

Researchers have worked on developing Maxwell's ideas into an instrument to transmit electromagnetic waves through the air. In 1894, Guglielmo Marconi was credited with building the first commercial radio system. It transmitted telegraph signals by radio waves. There was no sound until AM radio was invented in the early 1900s. Widespread AM broadcasting was not established until the 1920s, when vacuum tubes were developed.

Radar was the next technology to make use of high frequency electromagnetic waves. Beginning around World War I and continuing through World War II, independent researchers had to suspend working on their ideas, as development was guarded as a secret of war. During the 1930s and 1940s, radar was installed in battleships, submarines, and airplanes. This was often against the wishes of the commanding officers, not because of health effects but for tactical reasons.

One of the first uses in warfare was in 1941 in the battle between the Bismarck and the Hood. The commanding officers of each preferred to use their optical rangefinders to their newly installed radar. At that time, radar was unable to measure distance with the accuracy required for the main caliber guns.

By 1943, radar operating on microwave frequencies replaced older types of radar operating on VHF. The VHF band was later designated for TV.

Once upon a time... before there was cable or the Internet, there was VHF TV. To get good reception, a homeowner had to install an antenna on their roof.

What EMFs
Can Do to You

What EMFs Can Do to You

Health Effects Related to Electric Power

The laws of physics do not care how big or small voltages, currents, and charges are. If there is a charge on a particle, it wants to flow. To prevent it from flowing, something has to block or contain it. A charged particle is going to become frustrated if it can't flow. It may build energy and become volatile. Consider a battery. A battery will explode if the charger does not stop when it is full. Making the assumption that the body cannot be affected by electric and magnetic fields from electric power is illogical. The body is not insulated or protected from electricity.

One of the basic laws of electricity and magnetism is that an electric field will induce a magnetic field at 90 degrees to it, and vice versa. These fields affect currents in the body used by cells and nerves to communicate.

What EMFs Can Do to You

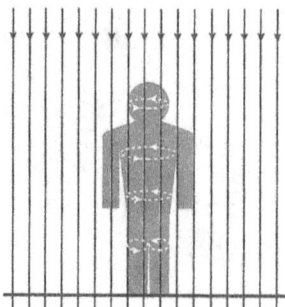

A person standing in a magnetic field showing the induced electric current on the body. The dashed circles are the induced electric fields.

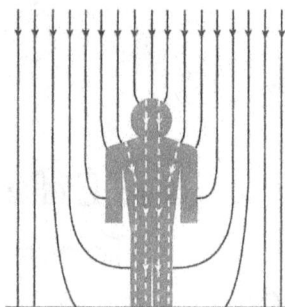

A person standing in a electric field showing the induced magnetic fields on the body. The dashed lines are the induced magnetic fields.

Non-carcinogenic health effects associated with EMFs include the following (DOE/EEE-0040, p. 23):

- changes in the function of cells and tissues

- decreases in the hormone melatonin

- alterations in the immune system

- changes in biorhythms

- changes in brain activity and heart rate

EMFs have also been associated with depression, insomnia, infertility, headaches, birth defects, memory loss, cramps, and buzzing and ringing in the ears. Although researchers have suggested that EMFs do not cause cancer, they concede that EMFs promote or allow existing cancers to develop. It is thought that EMFs cause a disruption in normal cell activity and interfere with or suppress the immune system and normal bodily functions. The following organizations have listed EMFs from electric power as a potential heath hazard or possible carcinogen:

- Environmental Protection Agency (EPA), 1995.

- California Department of Health Services (CDHS), 2001.

- World Health Organization (WHO), 2001.

- National Institute for Environmental Health Sciences (NIEHS), 1998.

- International Agency for Research on Cancer (IARC), 2001.

Possible cancers include childhood leukemia, lymphoma, and brain cancer.

Cows who were canaries

Numerous cases have been reported of dairy cows becoming sick and experiencing decreased milk production due to what is referred to as "stray voltage." Stray voltage is the result of the electrical power distribution system being connected to the ground.

What EMFs Can Do to You

According to "A Brief Chronology of Stray Voltage Developments on Dairy Farms in Wisconsin," going back to the 1940s, there was nothing regarding stray voltage in the literature until dairy farms became electrified. Milk lines used to be glass, not metal. In the 1960s and 1970s, farms became electrified, steel pipes and metal stanchions were added, and rebar was added to concrete. When concrete becomes wet, it's a low resistance path for current to return to electric power substations.

Farmers began to complain in the 1960s, but authorities blamed the cows' symptoms on faulty wiring rather than ground currents. As mastitis, an inflammation of breast tissue due to infection, became prevalent, it was treated with antibiotics and blamed on poor farming practices.

In the 1970s, the grounding issue was finally "discovered." The solution was to drive more grounding rods into the ground at the electrical panels on the farms, which only served to make the situation worse.

In the 1980s and 1990s, rather than making changes to the electrical code, which would eliminate ground currents, the government and the electric companies decided to create exposure levels at which cows could be expected to tolerate the effects. The acceptable voltage was set at 1 V. (Although zero is safest, one seems a common number. The EPA, for example, does not consider something to contain asbestos until it contains at least 1% asbestos fibers, even though there is no safe level of exposure to asbestos fibers. One percent is based on economics and preventing panic; not health.)

The ultimate solution for a farmer who had money was to install an isolation transformer. Inside an isolation transformer, the wires from a farm are not connected to the wires from the electric company. Isolation is accomplished using electromagnetic devices and a principle of physics called magnetic flux.

DOE/EEE-0040. *Questions and Answers About EMF, Electric and Magnetic Fields Associated with the Use of Electric Power*. National Institute of Environmental Health Services and U.S. Department of Energy, U.S. Government Printing Office. Washington, D.C., January, 1995.

"A Brief Chronology of Stray Voltage Developments on Dairy Farms in Wisconsin, The Perspective of a Dairy Farmer and an Electrical Engineer." October 2, 1998. Spark Burmaster.

What EMFs Can Do to You

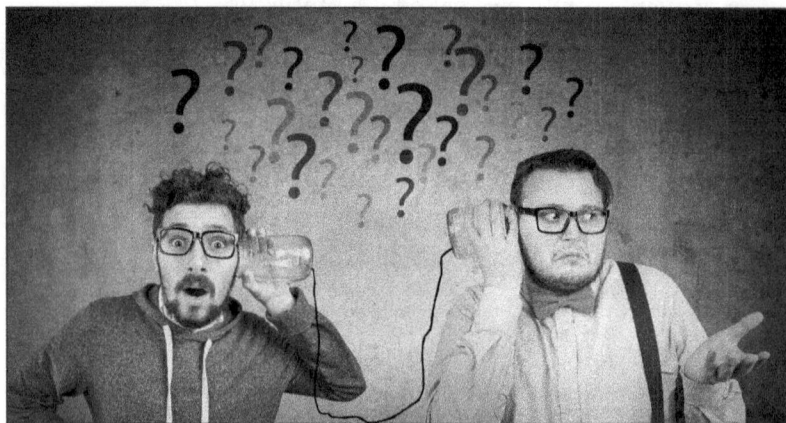

Health Effects Related to Radio Frequency Waves

Unless enough power is present to raise the temperature of the cells in your body by one degree, the FCC and other government organizations do not consider high frequency emissions to be harmful. Heating is considered the only possible biological effect.

People have claimed to be affected when levels are lower than those required to heat them. Some report being affected by power so negligible that even a meter can miss such a weak signal if it is not set correctly. Unlike the electric and magnetic fields from power lines, which have a large volt per meter (V/m) fields, the V/m levels emitted from a cell phone, WiFi, and other wireless technologies are low.

If not the amount of power, what is bothering people who claim to be sensitive to this type of EMF? Perhaps the data modulated on the carrier frequency is a factor. The frequencies that WiFi, cell phones, and other wireless technologies operate on are called carrier frequencies. They act like waves in the ocean. More than one user rides the same wave at the same time.

When you make a call or use the internet, data is modulated on top of the carrier waves. Your data is like a surfer riding the wave. There are others surfers riding the same wave. Perhaps the body is sensitive to the modulated data—thousands of voices and emails vibrating though every cell.

Only the strength, the size of a wave, is measured with a meter. The data are not measured. If you want to see or hear the data, pick up the phone or check your email.

Not a bright idea

Near a person's head may be the worst place to put a cell phone or bluetooth device. Eyes are sensitive to electro-magnetic radiation, which makes sense since light is an EMF. Numerous case studies concerning damage to eyes are presented in *Occupational Health: The Soldier and the Industrial Base*. Effects include retinal and corneal damage and degraded color and night vision.

Retinal injury is considered photochemical, which means molecular bonds are broken. The rule of exposure states that the same dose produces the same effect delivered at a high power for a short time or a low power for a long time.

The cells and bones of the ear are designed to amplify frequencies. The resonate frequency of the ear canal is 2.5-3.5 GHz. This provides a boost of energy to sources that enter it. Hearing loss is therefore found slightly higher, in the 3-4 GHz range.

We tend to think amplification is limited to what we can hear. Audiophiles use speakers capable of frequencies above and below what can be heard. It's thought that the body uses frequencies for perception, even those that we cannot hear.

You don't need to get burned to feel heat

Cells can be heated from exposure to (RFs) without the body temperature rising. The body regulates its temperature and expels heat during normal metabolism. You can be heated from the inside out. An RF burn is deeper than an ordinary burn. A normal burn is continuous on the skin; however, a burn from an RF depends on the frequency. Microwaves, for example, excite water molecules. Tissues with higher water content (i.e., skin and muscle) are affected more than fat.

Exposure limits are based on body weight. Women, children, the elderly, and those who are immunocompromised may have different compositions of tissue, muscle-to-fat ratios, and water content.

The resonator effect

The *Textbook of Military Medicine* says coupling energy from an RF source at the resonant frequency of the body will increase the energy the body is exposed to three-to-five-fold. The text does not state at which frequencies the body operates on. Bacteria, viruses, and other organisms operate at different frequencies. It's possible that some organisms and cells are supported while others are suppressed.

Changing limits

Official Permissible Exposure Limits (PELs) were set to 10-fold greater than the levels at which thermal effects were expected. In recent years, some PELs went up, not down. Increasing power levels due to new sources make compliance difficult. Higher exposure for determined lengths of time may be permitted under certain circumstances based on the frequency and the worker or military occupation. Exposure above a limit may be acceptable if it does not exceed a certain amount of time.

Occupational Health: The Soldier and the Industrial Base. Textbook of Military Medicine Part III Disease and the Environment. Chapter 15, Non-Ionizing Radiation. U.S. Army, Office of the Surgeon General,1993.

Bluetooth® is a registered trademark and is used here as generic descriptor similar to coke, xerox, and kleenex.

Meters

Meters

All-in-one Meters

All-in-one type meters that measure more than one type of field are not recommended. It may seem that you are saving money, but these are not as accurate or useful as meters that measure only one type of field. An all-in-one meter could lead you to make the wrong conclusion regarding the level of the fields present and the effectiveness of the changes you make to reduce them.

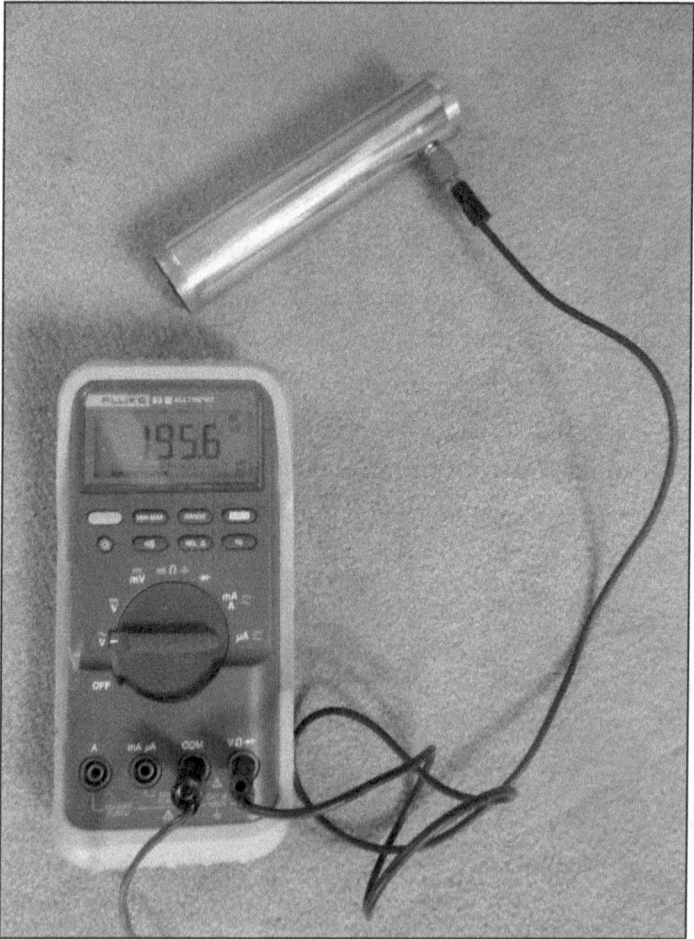

A multimeter being used to measure body voltage. One cord from the meter is connected to a metal object (the pipe). The pipe is held by the person reading their body voltage. One cord from the meter is connected to a metal tent stake pounded into the ground outside.

Meters for Measuring Electric Fields

To accurately measure electric fields requires a meter that costs around US $10,000. This meter has a fiber-optic cable that connects it to a probe on a tripod. The tripod is placed where a reading is desired, and the user steps back. If this is not done, the person taking the measurements influences the field, resulting in a less accurate reading.

You don't need that kind of meter to determine what the biggest sources of electric fields in your home are, or the effectiveness of the changes you make to reduce them. I suggest the German Bau-biologie method. A multimeter, the same as that used by electricians, can be used to measure voltage on the body. This is called "body voltage."

The following supplies are required:

- a digital, auto-range multimeter capable of reading as low as 1 mV (millivolt)

Meters

- a 36-inch, banana-to-banana test lead cable. The standard size is 4 mm. Make sure the multimeter you have accepts 4 mm connectors.

- a 4mm banana socket binding post nut banana plug jack connector

- something metal to hold—a short piece of a metal drainpipe from the plumbing department works

- 50 feet of lamp cord

- two banana plug connectors, 4 mm or the size that fits your multimeter

- an alligator clip

- a tent stake or metal spike

The multimeter must be capable of reading as low as 1 mV. A multimeter of this type should cost US $100-$400. A cheaper multimeter may only measure higher voltages.

Lamp cord can be purchased off a spool at a hardware store. Get 50 feet and split the cord in half. You only need one wire. Discard the other half. On the wire you keep, cut back 1/4 inch of the plastic and install a banana connector on each end. Place the tent stake or metal spike outside into the ground outside. It should go into soil. Connect one end of the 50-foot cord to the tent stake using the alligator clip by plugging the banana connector into the alligator clip. Connect the other end to the meter input labeled COM (common) using the banana connector.

An alligator clip is used to connect the cord to the metal stake in the ground outside.

Use a small drill bit to drill a hole in the metal pipe and install the banana socket nut to the pipe. Use the 36-inch banana-to-banana cable to connect the pipe to the meter input labeled V (Volts).

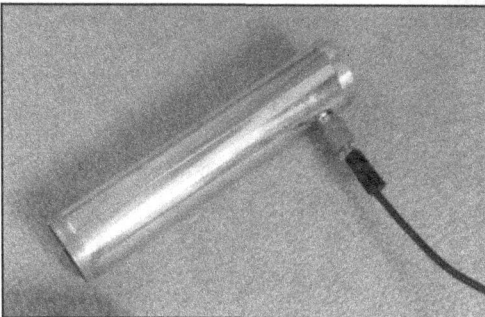

Install a banana socket nut on the pipe to connect it to the banana-to-banana, test lead cable.

Set the meter to ACV (Alternating Current/Volts). Hold the metal pipe and observe the voltage. That's the voltage induced on your body from the electric fields in the house.

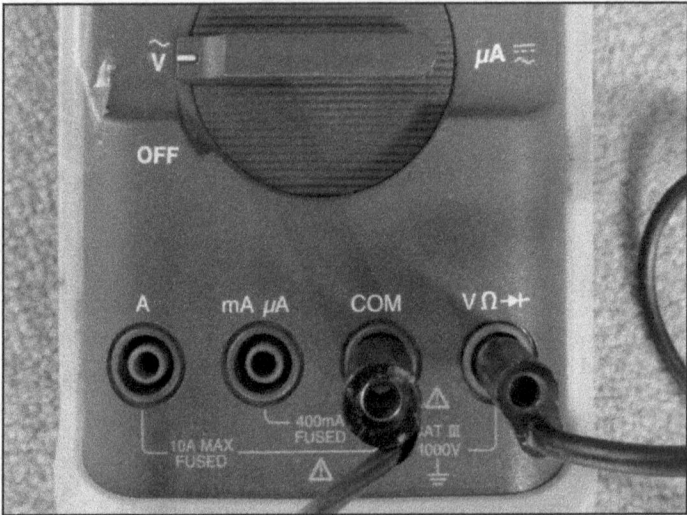

Connect the cord from the tent stake to the COM input. COM is the same as the ground. Connect what is held (the metal pipe) to the input on the meter labeled V (Volts). Set the meter to read AC volts. On this meter, the symbol for AC volts is a V with a squiggly line on top indicating that the current is alternating like a wave instead of constant like a straight line.

If you do not want to build a body voltage meter, meters may be found on-line at stores such as LessEMF.com. For $130, you can buy the kit: the body voltage meter, the ground stake, and 50 feet of cord. Beware of eBay listings that say "Body voltage kits." The parts won't work unless the multimeter you have is capable of reading as low as 1 mV. The cord provided is made to plug into an electrical outlet; body voltage readings should be taken by plugging the cord into a stake in the ground outside. The metal piece it comes with to hold might be worth the price. It will save time getting a piece of metal pipe to hold. The piece of metal to hold comes with a connector pre-installed to plug into the meter.

Meters for Measuring Magnetic Fields

A gauss meter is used to measure magnetic fields. Carl Friedrich Gauss (1777-1855) was a German mathematician and physicist who developed the method for measuring magnetic fields. The gauss is the unit of measurement used for magnetic fields.

I use an F.W. Bell Model 4180 that is digital and tri-axial, meaning it measures all three directions at once. Founded in 1944, F.W. Bell is the best in the world at designing and manufacturing gauss meters.

In the early days

When I started measuring fields, I bought the cheapest meter available, which cost US $50. When I had enough money, I bought the F.W. Bell. Soon after, I received a call to measure power lines next to a house. The owner was concerned, as the power lines were 10 feet from the house and on the same side of the house as the kids' bedrooms.

Meters

I took both meters. As the client watched, I walked from the house towards the power lines. When I was nearly under them, I stopped and looked at the meters. They had different readings.

My client was a commercial builder and had concrete barriers, the type used for highway construction, running between the house and the power lines. We ducked behind the barriers to see if they shielded the magnetic fields from the power lines. The readings on the cheaper meter decreased; the readings on the F.W. Bell stayed the same.

When purchasing a meter, you want one that does not give incorrect low or high readings when it gets too close to a source. Concrete does not shield magnetic fields. The rebar might change the harmonic frequencies and the cheaper meter might be sensitive to harmonic distortion.

Meters for Measuring High Frequencies

As frequencies get higher (through the use of WiFi, cell phones, and smart meters), electric and magnetic fields become inseparable and are no longer measured. The total power present for a specified frequency is measured instead.

A cheap meter might not be helpful, and by cheap, I mean US $4,000. It might only display the total power for the range of frequencies it can measure. If, for example, a meter can measure 100 MHz (FM radio) to 3 GHz (WiFi), it will display the total power present from everything in that range. The user gets one reading combining every FM station, cell phone, WiFi, and bluetooth nearby, thus limiting the usefulness of the reading.

You might, for example, want to compare readings between rooms. It could be that the readings appear to be the same when only the total power is the same. The bulk of the reading in one room, for example, may be from WiFi. As you move to another room, the strength of the WiFi might decrease and the FM reception might increase, resulting in the same reading in both rooms.

One way to tell if a meter is not the most accurate is to approach a source such as the WiFi and go right up to it. If the meter goes berserk, it's overreacting. Thus, in the middle of the room, it might be under reacting.

To successfully measure and investigate sources of high frequency EMFs, you need a spectrum analyzer. Cheap meters exist that claim to be spectrum analyzers. I paid $3,000 for one and sold it after realizing the readings were not even close to being correct. The type of meter engineers use to take readings cost upwards of $30,000. Rent one.

An example of a good meter to rent is the NARDA SRM3000. With this one, you can selectively measure EMFs due to the 2.4 GHz band used by wireless devices such as WiFi and Bluetooth. Experiment with powering each of these off to see which ones have the strongest affect on the total. You might be surprised to find that a wireless keyboard or mouse is emitting a higher signal than the WiFi. You might find that the highest signal's source is a cordless phone. If so, you could reduce the radiation in your home by finding a phone that emits less power and switching to a corded phone for long calls. If the highest source is a cell tower nearby, you'll know it for sure, and you can try tinting the windows or make curtains out of fabrics that have shielding properties. You can't make these determinations with simpler, low-cost meters.

High Frequencies

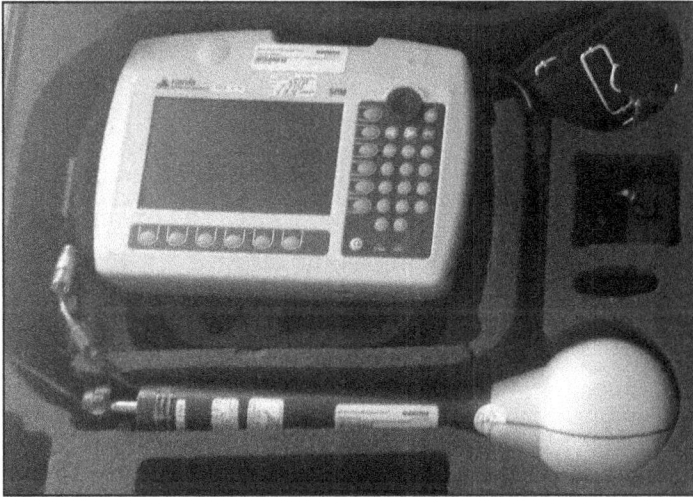

A NARDA 3000 high frequency spectrum analyzer. Narda Safety Test Solutions is the leading manufacturer of equipment for measuring high frequencies. If you need to do an evaluation that requires accurate measurements, consider renting one of their meters.

I discovered how inaccurate some meters are by testing them side by side. When I bought one for $3,000, I thought I had upgraded to something special. One day, I paid $1,500 to rent the $30,000 meter for a week. I was shocked at how inaccurate my $3,000 meter was.

Don't fret. In the section on getting rid of fields, there are suggestions for reducing the fields, even if you can't accurately measure them. Just don't call the cell phone company and complain about radiation. They will dispute your complaint with good cause.

Case study: Choosing a classroom

This is a case study of a classroom building where several vacant rooms were being considered for a new class. There was an administration office nearby, the source of the WiFi. The concern was a cell tower on the same side of the building as the classrooms. The tower could be seen in the distance.

Readings were taken outside, in the line of sight of the tower, and inside, in the potential classrooms. Two meters were used. The first was the $30,000 NARDA, and the second was a $4,000 meter. The cheaper meter underestimated the peak (maximum) readings. It might have been useful if there was a way to calculate the relationship between the readings it was giving and the actual ones. It was unpredictable and erratic, especially when it got close to a source.

The highest field in all the rooms *and* directly outside the building was from WiFi. After WiFi, the next strongest signal in Room 2 was TV and FM radio, which made sense because that room had a window on the same side of the building as the hills in the distance. After WiFi, the strongest signal in Room 4 was from cell phones *inside* the building. It was the phones inside, not the tower outside, that were the strongest sources of cell radiation. Consider what conclusions you would make if you only had the cheaper meter.

High Frequencies

Readings in uW/m² (microwatts per meter squared)				
	Meter 1		**Meter 2**	
Location	**Peak (Max)**	**RMS (Average)**	**Peak (Max)**	**Comments**
Outside	7.3	2.6	0.14	WiFi was the highest, followed by cellular.
Room 1	56	4.6	2	
Room 2	1,130	42	13	Close to WiFi in administration office.
Room 3	39	5	4	
Room 4	51	6.4	1.5	

Meter 1: A NARDA SRM-3000 with a 27 MHz-3 GHz antenna, costing approximately $30,000. It was set to search for the best measurement range, RBW on auto. As it simultaneously keeps track of the average and peak (maximum) readings, both were recorded.

Meter 2: A meter marketed as an analyzer with a broadband antenna, measuring 27 MHz-3.3 GHz, costing approximately $4,000. It was set to maximum range, peak hold, standard VWB, 0 dB, and Full. Readings were collected in a figure-eight pattern to account for all directions. The meter is not capable of measuring average and peaks simultaneously. The peak readings were measured.

Meters

Meters for Measuring Dirty Electricity

The meters sold to plug into an outlet to measure dirty electricity do not have units of measurement. The manufacturer designed them to provide a quick and easy way for a homeowner to check how much dirty electricity is present. To measure dirty electricity to an electrical standard requires an oscilloscope and an analyzer. However, it is technical and not something the average homeowner or electrician can do.

The good news is that a battery-operated AM radio can be used to perform the assessment. It needs to be battery-operated. Go outside and set it to a channel with no station and little static. Take it back inside. If it gets noisy—that's due to dirty electricity.

Using an AM radio in this way is similar to using a combustible gas meter to find gas leaks. A combustible gas detector has a sensor that must be calibrated outside and in a clean environment. It emits an audio signal that beeps more rapidly in the presence of gas. The sensor does not detect gas. Rather, there is a change in resistance in the circuit when gas is present.

How to Take Readings

How to Take Readings

Measuring Electric Fields

Readings are taken using the Bau-biologie, body voltage method. The focus should be in bedrooms and where time is spent, such as at a desk in a home office or a study area. While it is impossible and impractical to avoid electricity, it is desirable to have a sleeping area where fields are as low as possible.

Make a data sheet to remember what you do, in which rooms the readings were taken, and what they are.

Electric Field Readings	
Body Voltage milli-volts (mV)	Comments

Arms and legs act as antennas. Readings will be different when laying flat compared to standing. Have the person being measured lay on the bed as if sleeping. Depending on the materials used to construct the bed (i.e., metal, wood, and/or plastic), readings taken while lying on the bed will differ from those taken while standing or sitting on the floor. Do not stand close to the person being measured, as your body will affect the reading.

Make sure the cable is securely connected to the tent stake outside. Make sure the meter is set to AC, not DC. Take a reading.

Although the meter is set to volts, when a reading is lower than 1 V, it may switch to display in mV. One volt is equivalent to 1,000 mV.

If you get confused as to how something appeared to radically increase or decrease, check the units on the meter. It may have switched from mV to V.

Electric Fields

How to Take Readings

Measuring Magnetic Fields

If you own a meter, and it's digital and tri-axial as suggested in the chapter on meters, it should be as simple as
turning it on. Stand still when taking a reading.

Who can you call?

If you call them to ask, the electric company may send
someone to take measurements for free. No matter how
high or low the readings are, they will tell you there's no
cause for concern. What you get is a free reading but no
advice, and possibly not good measurements, as there are
conditions that effect readings the representative may
not be trained to consider.

Turn it on.

Magnetic fields are only present when current is flowing.
Unless things (e.g., lamps, lights, and WiFi) are turned
on, there won't be a magnetic field unless it comes from
power lines. Turn on all the lights and leave them on

until you have finished taking readings. The exception is when taking reading to assess if power lines are the source. The power inside should be turned off when checking, so you know the reading inside can only be from power lines outside.

Make a map.

Make a drawing of the property and floor plan of the house. Get two different colored pens. Use one for readings with the power off and the other for readings with the power on.

Don't worry about

- refrigerator

- microwave oven

- kitchen range

- window-mounted air-conditioners (unless a bed is nearby)

These are normally not close to where people spend a lot of time.

Don't worry about the following:

- computers

- office equipment

- TVs

Do not take readings by placing the meter next to one of these. Take readings where you sit—place the meter on the chair at a desk, on a couch, or on a bed. Take a reading at the head of a bed behind the wall of a refrigerator.

Worry about

- power lines

- the fuse box

- wiring errors

- current on metal water pipes

These are what to investigate if you find the readings to be higher than the recommended levels.

Take some readings

Take readings at each corner and in the center of each room. Take readings in living rooms, dining rooms, and areas people spend time sitting or sleeping. Place the meter on the couch and on the dining room table. Don't worry about bathrooms and laundry rooms. Don't worry about the kitchen. Kitchens always seem to have elevated magnetic fields. Often it's from the refrigerator, which may cycle on and off. New appliances have controllers that are always on. To check the level of the magnetic field, unplug the appliance and see if the field drops. If, for example, the dishwasher is constantly emitting a field, the solution is to install a switch to turn it off when it's not being used.

Gauss meter being used to take a reading on a pillow on a bed.

Do not take readings close to the walls. There are going to be fields next to light switches and the wires inside walls. Do you stand against the wall all day? Put the meter where you normally sit and spend time. Bedrooms are critical. Take readings at the head of your bed and on the pillow.

Measurements can change with the time of day and the demand for power. Take readings in the morning when people are getting ready for work and school, and again in the evening when people are home cooking dinner, watching TV, and running the airconditioner. Levels may be higher in the middle of the day if the power lines are serving businesses or manufacturing facilities.

A drawing of a house and property with readings when the power was off. The concern was power lines running next to the left side of the house and in the alley behind the backyard.

How to Take Readings

Measuring Dirty
Electricity

Get a battery-powered AM radio. Go outside and set it to
a station with no reception. Go back inside and walk
around the house. If you hear noise on the radio, it's
coming from noise on the electric wires. This is known
as dirty electricity.

Turn the power to the house off. Walk around inside
the house while listening for noise on the radio. If the
noise goes away when the power is shut off, the sources
of dirty electricity originate inside the house. See the following chapters on how to reduce it. If the noise persists
when the power is off, there may be an issue with the
power supplied by the electric company.

How to Take Readings

A NARDA SRM3000 on a tripod in a backyard being used to measure radiation from nearby cell phone towers and an airport radar.

Measuring High Frequency Fields

To be accurate, you must set parameters on the meter depending on the sources you want to measure. The settings for WiFi, radar, cell phones, AM and FM radio, and TV are different. The settings must be correct in order not to miss or overestimate fields. One of the difficult parts of measuring high frequencies is that even if you rent an expensive meter, it needs to be programmed with this information.

The following are possible settings. Cheaper meters may require additional settings. If your meter does not have a place to set these, it's not a spectrum analyzer.

Minimum and Maximum Frequencies indicate the range of frequencies each source operates. WiFi, for example, broadcasts on three ranges: 2.40-2.5 GHz, 5.1-5.3 GHz, and 5.7-5.8 GHz. Consumers know these as normal and high-speed internet.

Resolution bandwidth (RBW) determines how close two signals can be while still being resolved by the analyzer into two separate peaks. The narrower the

RBW, the more clearly the meter can see two close sig-nals as separate; 3 MHz is a good default. Depending on the source, 1 MHz or 5 MHz may be appropriate.

Video bandwidth (VBW) is set to full for weak sig-nals or between 1 and 3 MHz; 3 MHz is a good default.

Decide how long you are going to stand in one spot and take a measurement. If you are interested in the peak reading, you want to wait at least five minutes. The FCC may use five minutes to calculate an average expo-sure.

There are different antennas. The standard one cov-ers the range of frequencies of common concern: cell, WiFi, radio, TV, and radar. If you want to measure higher or lower frequencies (e.g., signals from satellite, military, radar, and 5G), you need to change the antenna.

Service	Fmin	Fmax	Distribution
TV Ch. 2-6	54.000 000 MHz	88.000 000 MHz	33.46 %
TV Ch. 14-69	470.000 000 MHz	806.000 000 MHz	22.55 %
PCS Broadband	1 850.000 000 MHz	1 990.000 000 MHz	12.22 %
FM Radio	88.000 000 MHz	108.000 000 MHz	10.26 %
TV Ch. 7-13	174.000 000 MHz	216.000 000 MHz	9.87 %
Paging	152.000 000 MHz	159.000 000 MHz	2.157 %
Land mobile&Ham	902.000 000 MHz	930.000 000 MHz	1.803 %
Cellular AMPS	824.000 000 MHz	849.000 000 MHz	1.623 %
Cellular AMPS	869.000 000 MHz	894.000 000 MHz	1.601 %
ESMR/Land mob.	849.000 000 MHz	869.000 000 MHz	1.335 %
Total			100.0 %

With a push of a button, a spectrum analyzer can display the total power each source is generating. In this example, the meter is set to sort, high to low, the percentage each source makes towards the total.

Smart meters

At the time of writing, smart
meters were known to operate
between 902 and 928 MHz, and
repeaters at 1.9 GHz. If your
meter requires a setting, a sug-
gestion is a sweep time of 33
microseconds instead of the
normal 0.5 microseconds. Hold
the meter two or three feet in
front of the smart meter, not in
contact with it. The signal

An electric meter that is a
smart-meter.

strength may be low. Set the meter to detect weak signals
and a narrow bandwidth so that the meter is not over-
loaded by higher signals such as WiFi. It's helpful to ask
the utility company for the exact frequencies and pro-
gram the meter for them. When finished, do a full
spectrum analysis to see if the smart meter uses frequencies the utility company is not aware of.

If you are concerned about potential health effects, opt
out of having a smart meter. If that's not possible, consider shielding it. Shielding won't eliminate the field, and
it may not enclose the entire meter. If it did, the smart
meter would cease to operate, and the utility company
would come to investigate. It may only shield the back-
side, facing the interior. This is the same as installing
shielding in a phone case, which prevents direct trans-

mission towards the head. It does not change the level detected a short distance away. If it did, the phone would not work.

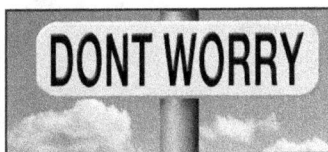

If you do not have a spectrum analyzer, don't worry.
Follow the recommendations in later chapters to make
your indoor environment healthy.

How to Take Readings

Measuring DC Magnetic Fields

W.F. Bell sells the best meter for measuring DC (Earth's) magnetic field. The probe is directional. Thus, unless it's pointed down, you won't get a peak reading.

Using the meter, walk around the house and make a note of any variations in the readings. Metal causes readings to shift. Concrete may affect readings, depending on the aggregate and the rebar. Metal bed springs, metal bed frames, and metal studs used to frame walls, can alter the field. Batteries used to store power for solar or wind can increase readings within a few feet.

Explore outside. DC gauss meters can be used to check for geomagnetic changes caused by underground gas and water pipes, underground water, minerals, and sink holes. It might be fun to compare the readings you get to an assessment performed by a well company looking for a place to drill a well.

What the Readings Mean

What the Readings Mean

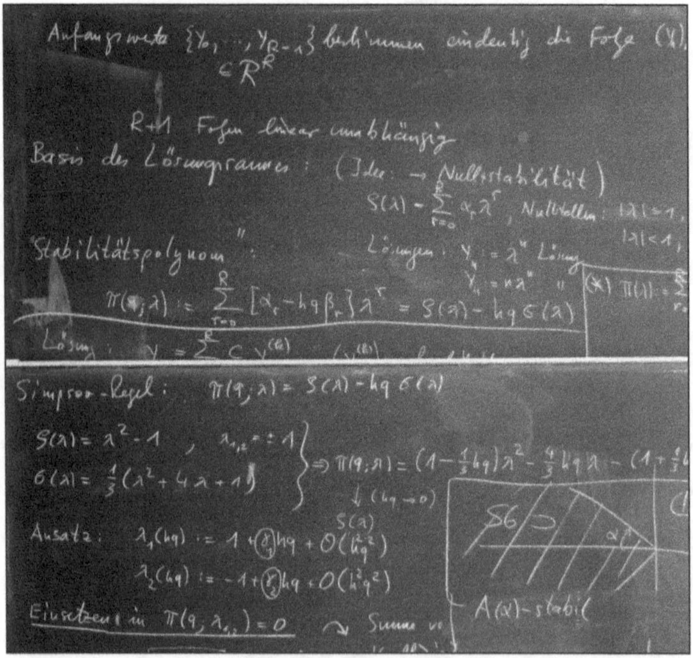

Don't Get Caught Up in the Numbers

There is limited information about how non-heating and non-carcinogenic levels of electromagnetic fields affect us. The numbers obtained while taking readings may not be as useful as seeing how low you can go. Nature is the low number on the yardstick. Avoid extremes. When practical, make changes to lower your exposure. Sometimes, the nice part about taking readings is knowing you're in a safe, healthy environment.

What the Readings Mean

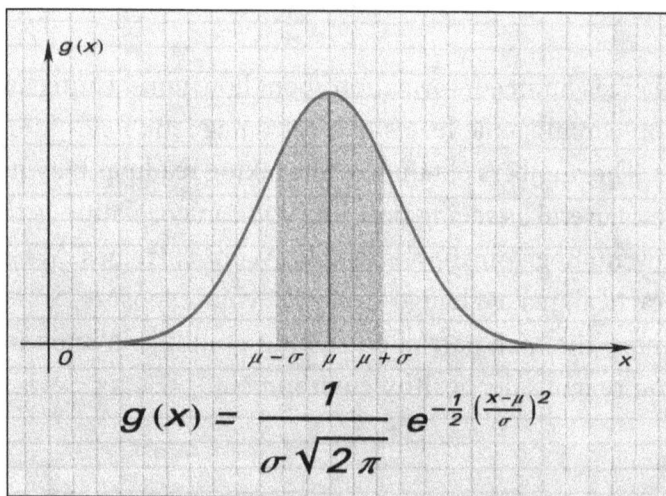

$$g(x) = \frac{1}{\sigma\sqrt{2\pi}}\, e^{-\frac{1}{2}\left(\frac{x-\mu}{\sigma}\right)^2}$$

Normal Levels

Use the values here as guidelines. They are not based on studies. They are relative to a few things: first, what's normal in nature (the low end); second, what's normal (average) in homes; and third, what might be considered high. There may not be a linear dose to exposure. Once above a certain level, it's hard to say how much worse, if any, higher levels may affect health.

The following guidelines for assessing levels of electric and magnetic fields are taken from the *Standard of Building Biology Testing Methods* by the Institut für Baubiologie and Ökologie, Neubeuern, Germany. These are based on nature as the yardstick. The lowest level might only be obtained by living outside in a tent in the woods. High (extreme) levels are based on what might be found in houses that measure high compared to others. These levels are intended to help make decisions. Prudent reduction, not complete elimination, is the goal.

What the Readings Mean

Electric Fields, Body Voltage	
Millivolts (mV)	**Concern**
Less than 10	None
10 - 100	Small
100 - 1,000	Strong
Greater than 1,000	Extreme

Electric Fields

You'll be surprised to find that no matter how clean a house is electrically, there will be some voltage on your body from the house wiring. Although the Bau-biologie level for extreme is 1,000 millivolts (mV), in most homes, the readings are between 1,000 and 2,000 mV (2-3 V). This is because most beds are against a wall with an outlet and wires hidden in it.

You can reduce the readings following the suggestions given in later chapters for getting rid of fields, which include unplugging things next to the bed, and, if necessary, moving the bed or sleeping in a different room.

What the Readings Mean

Magnetic Fields from Electric Power	
Milligauss (mG)	**Concern**
Less than 0.2	None
0.2 - 1.0	Small
1.1 - 5	Strong
Greater than 5	Extreme

Magnetic Fields

An Electric Power Research Institute study of 992 homes found the following:

  50% of homes had readings of 0.6 mG or lower,
  90% of homes had readings of 2.1 mG or lower,
  and only 6% had readings greater than 3 mG.

If the magnetic field increases by more than 0.5 mG when the power is turned on compared to when the power is off, some reduction could be achieved, as will be discussed in later chapters. If, with the power on, the reading is 0.5 mG or lower, it may be difficult to make further reductions.

Levels greater than approximately 2 mG have been associated with cancer, and in my opinion, readings of greater than 2 mG are higher than normal.

Source: Questions and Answers About EMF, 1995. Based on data in the table with credit to Zaffanella.

Voltage Distortion Limits According to Standard IEEE 519		
Buss Voltage (V) at the Point of Common Coupling (PCC)	Individual Harmonic Distortion %	Total Harmonic Distortion (THD) %
V ≤ 1 kV	5.0	8.0
1 kV < V ≤ 69 kV	3.0	5.0
69 kV < V ≤ 161 kV	1.5	2.5
161 kV < V	1.0	1.5

Based on the 2014 IEEE-519 *Standard Practices and Requirements for Harmonic Control in Electrical Power Systems*, page 6, Table 1. The Point of Common Coupling (PCC) is usually the service transformer. The electric company doesn't want harmonic current above a certain level to be impressed upon the neighborhood electrical distribution wires from a connected customer.

Dirty Electricity

Standard EEE 519 prohibits over 3%-5% total noise on the feed from the electric company. You may object and request that the electric company lowers it. First, you need an electrical engineer to measure it. The technique is technical and beyond the scope of this book. However, if you follow the recommendations provided in the later chapters, you can get rid of 90% of the dirty electricity in your home without getting the electric company involved.

....

EEE (Electric Equipment & Engineering Company) is the world's largest technical professional trade organization.

What the Readings Mean

High Frequency Fields

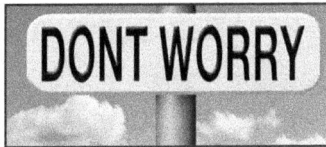

If you are overwhelmed by numbers, skip ahead.

The FCC has limits for Maximum Permissible Exposure (MPE). Exposure limits are calculated for a given frequency on the assumption that the only harmful effect is heating when tissue warms. The limit is 200 uW/cm² for frequencies of 30-300 MHz. The limit increases to 1,000 uW/cm² for frequencies between 300 MHz and 1,500 MHz. The limit for frequencies greater than 1,500 MHz (1.5 GHz) is 1,000 uW/cm². **In a residential environment, it's unlikely for a level of radiation to exceed these limits.** In the earlier case study, the average reading outside in the direction of a cell tower was 2.6 uW/m². The FCC maximum exposure limit is 1,000 uW/cm², which is much greater.

What the Readings Mean

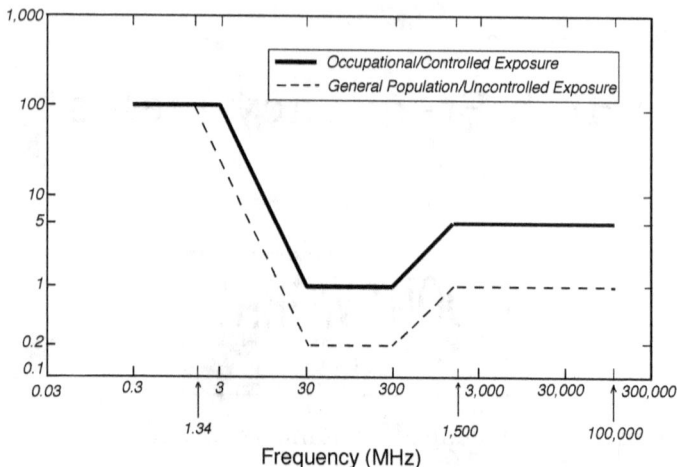

Figure 1. FCC Limits for Maximum Permissible Exposure (MPE)
Plane-wave Equivalent Power Density

Excerpt from *Evaluating Compliance with FCC Guidelines for Human Exposure to Radio Frequency Electromagnetic Fields.* **Federal Communications Commission Office of Engineering and Technology, August 1997.**

Recognizing there may be biological effects other than heating, some cities established stricter, more precautionary limits. An example is Salzburg, Austria, which set a precautionary limit of 1 mW/m². Unlike the FCC, which bases limits on average readings, precautionary limits are based on the peak, maximum reading observed during the measurement period. In the previously mentioned case, which assessed a cell tower, the peak reading was 7.2 uW/m². This level is below the precautionary limit of 1,000 uW/m².

Organizations and countries with precautionary limits set them arbitrarily. These are set to levels below FCC limits, but no one knows what level does not cause biological effects. There may be no safe level. Therefore, I do not consider precautionary limits to be useful.

The general public misuses precautionary limits. Many use meters that display only the total power present. A reading might appear to be high because the meter is displaying the total for all the frequencies present. Limits are applicable to individual frequencies. You must read the power emitted by each frequency (source) one at a time and then add up the total. Unless a $30,000 meter is rented, meters are likely to miss or over estimate peaks, causing people to worry more than they should.

While not useful for checking if exposure limits have been exceeded, readings can be useful to evaluate how much reduction was achieved by making changes. Maybe you can't live without WiFi, but it can be moved from the bedroom to the living room. An ethernet cable could be connected from the router to computers instead of using WiFi. Compare readings in these scenarios.

Readings can be used to make comparisons in different locations. It may be helpful to take readings in the home and compare them with the office, school, or a home you are considering purchasing.

What the Readings Mean

Exposure limits for cell phones

The FCC adopted guidelines for mobile devices in 1996. The limits are based on data that shows the body absorbs more energy at some frequencies than others. The strictest limits were set for frequencies of 30-300 MHz (FM radio and TV). At other frequencies, the body's absorption was thought to be lower, and higher exposure limits were set.

Exposure limits are based on the average exposure, not the peak, or maximum, level during the measurement. This as saying you cannot get burned by touching something hot only for a second because the average temperature of your body was lower. These limits apply to power densities "spatially averaged over the body dimensions." You might ask, "Whose body? Whose dimensions?"

Exposure limits depend on how the device is used. Hand-held phones must comply with limits for Specific Absorption Rates (SARs). A SAR is a measure of the rate at which energy is absorbed by the body when exposed to an RF. The maximum SAR permitted for mobile phones was defined as 1.6 watt/kg (watts per kilogram) as averaged over any one gram of tissue, defined as a tissue volume in the shape of a cube. What in the heck does that mean, and how does it translate to a real human? The manufacturer calculates the SAR, not the FCC. Exposure readings are not taken using the same methods as those used for evaluating exposure from transmission towers.

According to FCC guidelines, if we start using products operating at frequencies higher than 6 GHz, the SAR method is not appropriate. Power density needs to be measured, as is done for evaluating antennas on towers. The question arises as to what distance from a device a measurement should be taken. The ANSI/IEEE standard specifies 20 cm, almost 8 inches! The FCC suggests 5 cm (2 inches).

Next generation wireless

Technologies exist that could be used to lower the radiation cell phones emit. A cell phone only needs 10% or less of the power it's broadcasting to communicate with a tower. For example, suppose we are in a restaurant with 20 people and 5 want to make a call. Instead of using five times the radiation, why not piggyback the calls onto one person's phone? That person will not be exposed to more radiation than normal, and it would reduce the radiation to which we are all exposed.

Unfortunately, no matter what improvements are made, people are going to use the excess capacity. Instead of being happy the total radiation is lower, they are going to send more pictures and videos until the service becomes slow and requires an upgrade.

It's the same with electricity. Solar and wind were

supposed to reduce the amount of coal and gas being burned, but people use the power that's made available. If you want to do something good for the environment, use less electricity. If you are concerned about radiation from wireless products, reduce how often you use them or don't buy them.

Determining Why
a Field is High

Why a Field is High

Electric Fields

The following are common reasons for elevated electric fields:

- unshielded cords from lamps, clocks, and phone chargers next to the bed

- wires inside the walls next to and behind the bed

- wires running in the floor or ceiling

Which of these has the most significant effect on the total electric field is determined by measuring body voltage.

Unshielded lamp cords and wires

Wires and extension cords are big culprits. These bring fields from wires in the walls out into the bedroom, next to the bed, and closer to you. Unplug or move as many things as possible that are near the bed. Turning them off is not sufficient. This includes lamps, alarm clocks, and phone chargers. This is a two-person job. One person should lie on the bed holding and watching the meter for readings while the helper unplugs things. Remember to keep notes of what you do and what the readings are. If you unplug something and the reading changes significantly, this item/object should be permanently unplugged or moved away from the bed. Typically, 3-6 feet is sufficient.

Wires inside walls next to the bed
and wires running in the floor or ceiling

Locate the fuse box. Turn off each breaker one at a time, and record the body voltage. Walkie-talkies or phones are useful for communication between the person at the fuse box and the person lying on the bed holding the meter. You should see a big drop when you come to the breaker affecting the bedroom. If you do not see a drop, there may be wires running under or over the bedroom and circuits still on, having an effect on the bedroom. Leave the bedroom breaker off, and continue shutting other breakers off one at a time while watching the meter. Sometimes, one or two breakers in addition to the one serving the bedroom must be shut off to make a difference.

Compare the readings with the breakers off to those obtained by unplugging a lamp or clock radio next to the bed. You might conclude that it's not worth shutting the breakers off at night. (There's an easier solution in the chapter on getting rid of fields.)

You may discover there is power to the refrigerator or electric stove in the wall behind the bed and shutting off the power to the bedroom did not make a difference. If the appliance cannot be shut off, try moving the bed.

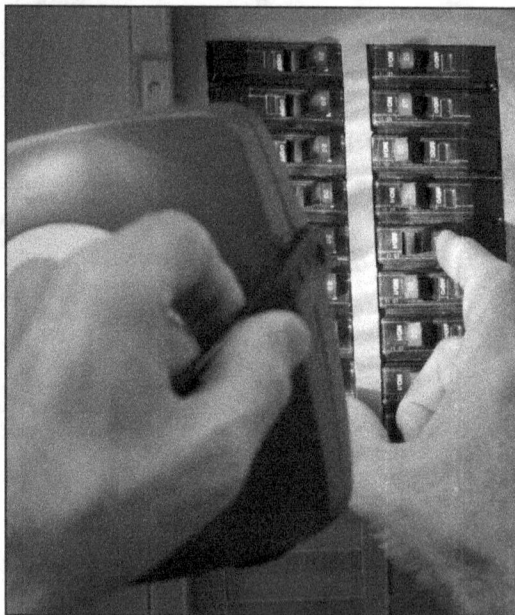

For this first part, turn off all the power.

When in doubt— STOP and call an electrician.

Magnetic Fields: Part I

Before investigating the levels of magnetic fields caused by wiring errors, first make sure the fields are not coming from outside. Take readings with the power off. The most common reasons for elevated magnetic fields when the power is off are:

- power lines (under and aboveground)

- current on metal water pipes

- current on cable TV and phone grounds

Check for power lines

Shut the power off. Go outside. Using a gauss meter, walk the perimeter of the property, taking readings at each corner and side. If you get high readings outside, go back inside and take readings with the power off. Record these on the drawing of the floor plan. If power lines are the source, you might have reached the lowest level possible.

Underground power lines can be significant. If the readings are high inside with the power off, go outside and watch the meter while walking towards the street. Check along the sidewalk and curb. If the readings increase near the street, the source is underground power lines. There is no way to shield them.

If you have a transformer nearby, its field should decrease quickly with distance. Take a reading near the transformer and slowly walk away from it. The field should dissipate within six to ten feet. If it does not, walk back towards the transformer, this time in the direction of the street or sidewalk. If the reading increases, the issue is unbalanced current on power lines underground. You can complain to the electric company. Chances are they won't change anything.

If you have high readings inside with the power turned off and power lines are not the source, these are the possible reasons:

1. current on metal water pipes
2. current on the coaxial cable for TV or internet
3. current on phone wiring

Check metal water pipes for fields

You should check these if you live in a house with metal water pipes between the house and the street. Place the meter next to the water pipe where the pipe enters the house. If the reading increases, it's because the city uses metal water pipes, and current is flowing in and out of your house on the water pipes. The solution is explained in the chapters on getting rid of fields.

Water main inside a house next to the water heater.

Water main near the street.

Check for current on coaxial cables and the phone ground wire

Cable TV and phone companies often connect their ground wires to the ground used by the electric company to ground the fuse box. Go outside and find the cable and phone boxes. Follow the ground wires (bare copper wires coming out of them). They may be connected to the big metal conduit to which the electric company attaches the meter. Place the gauss meter against the black coaxial cable and the phone ground wires. I know I said not to put a meter right up to wires. Ground wires and coaxial cables are exceptions. There should be no current on them.

Another way to check for current on a coaxial cable or phone ground wire is with a clamp-on ammeter. An electrician should have one. The jaws are wrapped around the wire or cable. The ammeter should read zero amps. Anything reading above 1/10 of an amp means there is current flowing on the ground wire. This is a problem. The solutions are provided in the chapters on getting rid of fields.

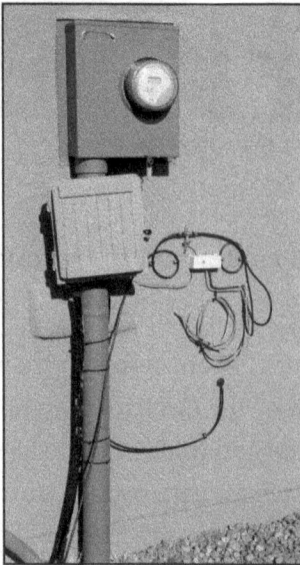

Phone and cable grounds connected to the metal pipe used by the electrical company.

Magnetic Fields: Part II

Turn the power back on. Go inside. Walk through the house and turn on all the lights. Devices such as computers, clock radios, and stereos may be plugged in and turned on or off as normal. Use the same drawing of the floor plan and a different colored ink to record readings with the power on.

Is there knob and tube wiring?

If you live in an old house and the fields are high everywhere, it may have knob-and-tube wiring. This style of wiring was prevalent from the 1940s up to the 1960s. Wires are suspended inside walls on insulated ceramic knobs. The only way to lower the field is to re-wire the house.

Why a Field is High

Check three-way switches

A three-way switch is one for which more than one switch may be used to turn a light off and on. Walk through the house and turn the light switches on and off, one at a time. If there is an increase when a light switch is turned on or off, this suggests a wiring error. Do not hold the meter next to the light switch. There is normally a field near a switch because the wires separate where they connect to it. If possible, get a helper to switch the lights on and off as you hold the meter and stand a few feet away from the switch, or set the meter on a table or the floor in the middle of the room as you turn the switches on and off.

If the meter is a few feet away from the switches, there should be NO change in the reading as a switch is turned on and off. A change means the current on the wires is not balanced. The likely reason is that the three-way switches were not wired with four-wire cable. Suggestions for fixing this are in the chapter on getting rid of fields.

Dirty Electricity

Use an AM, battery-operated radio. Go outside and set it to a channel with no reception and as little static as possible. Go back inside and find a lamp with an energy-saving light bulb. It may be a curly-Q fluorescent or light-emitting diode (LED) bulb. Turn the lamp off. Place the AM radio next to the bulb. Now turn the lamp on. The noise you will hear on the radio is being generated by the electrical ballast inside the bulb. An electrical ballast is a device that limits the amount of current that can flow. This creates dirty electricity.

The following create significant amounts of dirty electricity, and are responsible for 80%-90% of the dirty electricity in a home. Unplug or remove each of these and listen for the field (noise on the radio) decreasing:

- fluorescent lighting

- LED light bulbs

- Compact fluorescent light (CFL) bulbs

- low-voltage lighting with a transformer

- dimmer switches

To continue to investigate potential sources of dirty electricity, unplug each of the following as you listen for changes. The following have low-voltage transformers inside them:

- dishwashers

- microwave ovens

- cordless phones

- flat panel TVs

High Frequency Fields

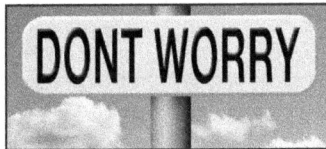

If you do not have a spectrum analyzer, follow the recommendations for getting rid of fields and do the best you can.

For an investigation to be useful, you need a meter with a spectrum analyzer. This shows how much power each source is emitting. If the highest source is your cordless phone, for example, you'll know to find a brand that emits less power or to switch to a corded phone. If the highest source is a cell tower outside, you'll know it for sure, and you can try tinting the windows or making curtains out of fabrics with shielding qualities.

With the meter set to a narrow bandwidth reading for the 2.4 GHz spectrum, try powering off the following and see which has the strongest affect on the total:

Why a Field is High

- wireless printers

- wireless stereo speakers

- WiFi

- computer keyboard and mouse

- wireless time machine backups

- cordless phones

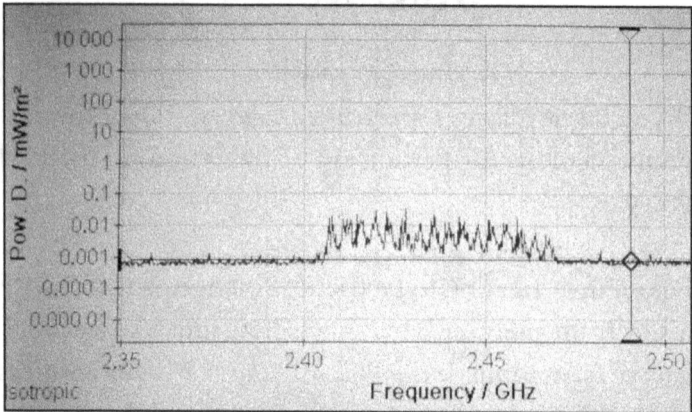

A spectrum analyzer. The peaks in the middle show where the activity is.

You might be surprised to find the keyboard or mouse for a computer is emitting a bigger signal than the WiFi:Find ways of eliminating big sources. Consider getting a printer without WiFi and turning the keyboard and mouse off when a computer is not being used.

High Frequency Fields

Getting Rid of Fields

Getting Rid of Fields

Getting Rid of Electric Fields

Unplug things near the bed

Unplug or move as many things away from the bed as possible, including lamps, alarm clocks, and phone chargers. Turning them off is not sufficient.

Move the bed

Some rooms have more wires in the walls than others. Sometimes, moving the bed makes a big difference. Try sleeping in other rooms in the house if you have alternative options. Here are some guidelines:

- TVs, refrigerators, and water heaters should not be on a wall opposite a bed.

- Electric baseboard heaters should be at least three to five feet from a bed.

- Transformers for 12 V lighting should not be on the ceiling of the room below a bedroom.

Change the cords on lamps

As soon as you plug something in—such as a lamp or light fixture—the unshielded cord brings an electric field out into the open. In bedrooms, living rooms, and where a lot of time is spent, replace the cords on lamps with shielded cords.

A shielded cord needs to be specially ordered. Belden makes the recommended cable. Use 18/2 for 10 amp (standard lighting) and 14/2 for 15 amp circuits (kitchens and garages). The 18/2 is thinner, making it easier to retrofit into light fixtures. An electrician can do this.

Shut the breakers off at night

Try shutting off the breakers before going to sleep. Some people report that the effect is like camping. Shut off the breakers to the bedrooms and to as many rooms as possible next to and around the bedrooms. Be careful not to shut off appliances such as the refrigerator or a security system. You will need a battery operated lamp and alarm clock.

Install a remote shut-off

To make it easier to shut the breakers off at night, a product called a Remote Shut Off Switch can be ordered from Save Living Technologies. The product has a remote control, like that on a car key, to shut off the circuit breakers. The circuits connected should include those for the bedrooms and the rooms next to, over, and under the bedrooms.

Install a grounding cloth bed sheet

A sleep shield or grounding sheet is a bed sheet made with conductive material snapped onto a cord that is plugged into a grounding stake outside. Since the sheet is connected to the ground, the sheet is at zero volts. People who sleep on them have lower body voltage readings.

When I first learned about EMFs, I suggested using ground cloth, or sleeping sheets, as a way to reduce electric fields. I no longer recommend them because some people have reported feeling unpleasant. However, others have reported positive affects; therefore, if you buy one and it works, keep it. If it does not, send it back. Never connect a sheet using the cord supplied by the manufacturer to the ground in an electrical outlet. The ground in an electric outlet is not at zero volts. Manufacturers sell cords to plug into electric outlets because it's easier than running the cord outside to a stake in the ground. Drill a 1/4-inch hole through an exterior wall, run the cord through it, and seal the hole with silicon caulk. Cable and phone companies do it all the time.

Sleeping on a ground sheet does not have the same

effect as shutting the power off. Shutting off the power eliminates the source of the fields; a ground sheet squashes the field, reducing it to zero at the sheet.

Case study

In the master bedroom, body voltage readings were taken on the bed. With the power turned on, the reading was 600 mV. When the breaker to the bedroom was shut off, the level dropped to 175 mV. The power was turned back on, and a grounding sheet was put on the bed and plugged into the ground in the electric outlet the way the manufacturer suggests. The level was 100 mV. The sheet was unplugged from the outlet and plugged into a tent stake outside. The level using the stake outside in the ground was 34 mV, the lowest reading.

Make your own grounding pads or sheets

You can make your own grounding pads or sheets. Buy a roll of Velostat®, a poly material impregnated with carbon. Use lamp cord, as explained in the chapter on how to make a meter read body voltage, to connect the Velostat® sheet to a tent stake in the ground outside.

Electric Fields

Buy an anti-static wrist strap

Workers in the electronics industry wear these to protect sensitive parts from static electricity as they work. Connect the anti-static wristband to a metal stake outside in the ground, and wear it when you sit at a desk for long periods of time.

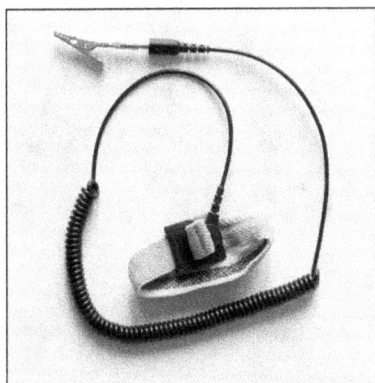

Getting Rid of Fields

Getting Rid of
Magnetic Fields

You may need a plumber

If you have a current on the water pipes and a high magnetic field when the power is turned off, it may be because the city has metal pipes and current is flowing on them into your house. This is both an EMF issue and a shock hazard. Call the electric company. It is an objectionable current. The electric company may come and check yours and your neighbors' service. New homes usually have plastic pipes. This is not an issue with plastic pipes.

Regardless of the electric company's assessment, to stop electricity from traveling on metal pipes, you must install a dielectric fitting on the main water pipe as it enters your home. Call a plumber. One or both, the electrician and plumber, may think you're crazy. After considering the issue, they will tell you it's a piece of cake.

Another way to stop current from flowing on the

water supply from the street is to cut into the metal pipe and install a three-foot section of plastic pipe underground, ten feet from your house.

"Serious shock incidents occur in the water industry at a rate of about one a day or 370 a year." *Distribution Systems*. Volume 90, Issue 7, July 1998.

Cable Guy

If the magnetic field is at the cable or phone wire, call the cable or the phone company. Ask them to disconnect their bond from the electric company's ground and install their own ground. It may take a few calls, because the first technician may say there's nothing wrong and question your meter. Ask to speak to someone in upper management who understands grounding. Some independent technicians (subcontractors) prefer to put in their own grounds—it reduces noise on the wires and callbacks for poor service.

When to call an electrician

Do not call an electrician until you can be clear and concise regarding what you want the electrician to do. Electricians don't have meters, and they charge by the hour. Never say, "I have high magnetic fields. Can you help me?" They may think you're crazy.

Magnetic Fields

Be prepared

The better prepared you are, the cheaper it may be. Before an electrician gets there, make a drawing of the location of every electrical outlet and light switch. Label it with which breaker turns on its power.

To identify which breakers control which outlets and receptacles, plug something into each outlet, one at a time. Get two lamps. Plug one lamp into the top receptacle on the outlet and the other into the bottom receptacle. Shut all the power off and turn on one breaker at a time until the lights on both lamps go on. Make a note on the map with the circuit breaker number. Make a special note of outlets where the lamp powered by the top receptacle is powered by a different breaker than the lamp on the bottom receptacle. Make a note of any outlets that are controlled by a switch on the wall. When this is the case, the neutral wires of the different breakers may have been accidentally wired together—something to ask the electrician to check.

Do the same with the light switches. Determine which circuit breaker controls each light switch. Having your map prepared will show where there are transitions of lights, outlets, and switches from one breaker to another. Problems often occur when there are switches on a wall where more than one breaker controls the switches inside it. When this occurs, ask the electrician to check that the neutral wires of the switches are not tied together.

What to say to an electrician

Talk to an electrician in terms they understand. They are familiar with the terms "unpaired wiring" and "unbalanced current." Initiate communication with something like this:

> There appears to be unbalanced current on wires in my home. This is causing an elevated magnetic field inside my house relative to outside. The issue appears to be that the wrong type of wiring was used to install three-way switches, or there is another type of wiring error causing net current on the wires.

The electrician may ask, "How do you know?"

Reply with the following:

> The field goes up when a light switch is turned on. As you know, when current flows, a magnetic field is created. If the cables were balanced, the magnetic field created by the return current would cancel the field created by the supply. The return current is not going back on the same circuit.

Magnetic Fields

The magnetic fields goes up only when I turn on things controlled by three-way switches. Therefore, I'd like you to check that the correct type of wiring was used to install the three-way switches. I suspect there is no traveler. If that's the problem, we may have to remove one of the switches. I'll tell you which ones I can live without.

If the correct wiring was used, it may be that the neutrals of the switches are connected to neutrals of a different circuit with the same wire nut inside a box. Another possibility is that a neutral wire is accidentally touching a ground wire inside a box. I've gotten as far as I can without taking the cover plates off to look.

General tip: An electrician should have an ammeter, which can be used to measure current on a wire or pipe. Any wire it's placed around should measure zero amps regardless of how much electricity is flowing. If the current flows back on the same set of wires, the return amps flow in the opposite direction. The wire (or pipe conduit) should have a balanced current (zero net amps). There should not be a current on a metal water or gas pipe. When there is net current (anything over 1/10 of an amp), keep it simple—ask the electrician to figure out why and to make repairs to correct it.

Things to Ask an Electrician to Check

Three-way switches

If it appears a set of three-way switches was wired incorrectly, it is likely that all of the three-way switches were wired with the wrong type of wire. They are supposed to be wired with four-wire cable. The four wires are: neutral, hot, ground, and *traveler*. The traveler must go to every switch that can be used to turn the lights on and off. As four-wire cable with a traveler is cumbersome to install, electricians may do what is called "picking-up a neutral." They grab a neutral from another light or circuit nearby and use it for the traveler. The lights work, but the current is unbalanced, and there is a magnetic field in the house where both wires go. The solution, other

137

than cutting holes in walls to run the correct type of wire, is to pick one of the switches to keep, and remove the others. Tell the electrician which switch you want to keep and which ones to disconnect.

Outlets powered by two breakers

These are outlets in which the top receptacle is powered by a different breaker than the bottom receptacle. These may be located where a light switch controls one of the receptacles to turn on a lamp, but the switch does not control the other receptacle. The neutrals may have been tied together with the same wire nut, or the tab on the outlet (inside the box) may not have been removed when the outlet was installed. Therefore, the neutrals are connected. Ask the electrician to check that the tab was removed and the neutrals are not connected with the same wire nut.

Subpanels

If it's a large house, there was an addition, or it was remodeled, a subpanel may have been added. At sub-

Ask the Electrician to Check

panels, have the electrician check that the neutral busbar (bus bar) is not connected to the ground busbar.

A busbar is a metal bar to which the wire coming from the main panel (fuse box) is connected. Circuit breakers snap into place on the busbar. Electrical current returns to the subpanel through the neutral (white) wires, connected to the neutral busbar. The main neutral wire going back to the main fuse box is connected to the neutral busbar. There is a screw inside the subpanel that connects the neutral and ground busbars. By default, the factory has the screw put in place since the neutral busbar is normally connected to the ground busbar at the main panel. They should not be connected in sub panels.

If the screw is not removed, current from the sub panel flows back to the main panel on the ground wire (and metal pipes). This creates an imbalanced current on all of the wires serviced by the subpanel and large magnetic fields. The electrician needs to remove the cover from the subpanel and look for the screw.

One ground is good; more is NOT better

If none of the above reduces a high magnetic field and it seems to be coming from inside and not from power lines, it may be that an electrician or someone working on the house added an extra ground. They may have connected the ground wire of a circuit they were working on to the metal pipes. According to the National

Electric Code, a ground wire shall only be connected to the pipes in a single location—at the main electric panel.

An electrician can check for this by clamping an ammeter around the metal pipes. If there is current on a pipe, it might be because an extra ground was added. Ask the electrician to find and disconnect it. To be safe, ask the electrician why the ground was added. There may be something else that needs to be fixed.

An extra ground added to connect an appliance to a gas pipe under the floor in a crawlspace. There was current on the gas pipe.

Appliances are often culprits. To fix the issue, where permitted, a dielectric union could be installed on gas and metal water pipes where they contact an electrical appliance such as a hot water heater or boiler.

When the fuse box is the source

If the only place the field is high is at the fuse box, it might not be because there is a wiring error. As wires come into the fuse box, they must separate so they can be connected to what are called busbars inside the fuse box. The field should drop quickly with distance from the fuse box. If it's too close to a bed or desk, do not attempt to reduce it by purchasing shielding; this is not likely to work. A better way to lower the field is to either move the fuse box to a place where you don't spend time or replace it.

There are types of fuse boxes in which the neutral bus runs the full length of each breaker column. The field cancellation is greater because the distance the hot and neutrals are separated inside the panel is smaller. Ask an electrician to order one. This type of fuse box allows the hot and neutral wires to stay together until the hot wires are connected to the breakers. A normal panel has only one neutral bus, and it's on one side of the panel. The greater the distance the wires are separated is, the larger the magnetic field will be.

Gangs of neutrals

It takes two to tango. This type of error requires two circuit breakers. Current from a set of wires served by one breaker is getting onto a wire powered by a different circuit breaker. Often if you find the first breaker, you'll find the second if you have a map that shows which outlets and switches are served by which circuit breakers.

Look for a set of switches on the wall where one switch is powered by a different breaker than the ones next to it. The neutrals for the switches should be kept separate. It's possible the electrician installing the switches tied the neutrals together with the same wire

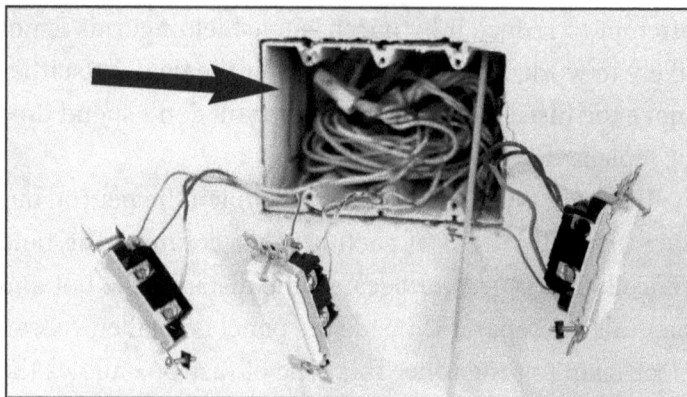

nut. This is called the ganging of neutral wires or "one big wire-nut syndrome" (see the image above). If you believe this is the problem, ask the electrician to remove the cover plate and check that the neutrals are not tied together.

A neutral wire touching a ground

This may be a difficult problem to locate and the last to consider. These account for a small percentage of problems. Here are some examples:

Example 1

The plastic jacket on a neutral wire was nicked when a canned light being installed was pushed into the ceiling. The exposed copper on the nicked neutral wire touched the metal can. As the ground wire was connected to the metal can to prevent shock, current on the neutral flowed onto the metal and then onto the ground wire back to the fuse box. Since rebar in the concrete was also connected to the ground wire at the main electric panel, the rebar in the floor had current flowing through it, creating large magnetic fields in places that there did not appear to be wiring.

Example 2

Stuffing wires into boxes. As they are not expected to have current, ground wires are not encased in a protective plastic coating. When a house is wired, the wires are stuffed into boxes. In so doing, the bare ground wire may touch the screw used to connect the neutral wire to the switch. When this happens, current gets on the ground wire and flows back to the fuse box on the ground wire instead of the neutral.

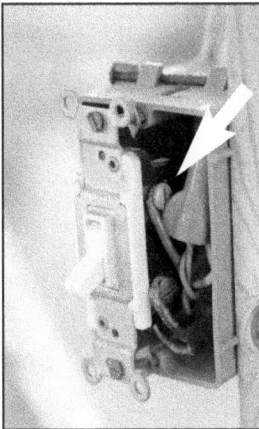

Example 3: This was under the cover of an electric baseboard heater. The plastic coating on the neutral wire was gouged when the cover was closed. Because the cover is metal, a ground wire was connected to it. The current on the neutral wire was able to get onto the metal cover and the ground wire. The current flowed back to the fuse box on the grounding system for the house, creating magnetic fields in various places throughout the house.

Getting Rid of Dirty Electricity

To get rid of dirty electricity, get rid of the following:

• fluorescent light bulbs

• LED light bulbs

• CFL (curly-Q, energy-saving light bulbs)

• low-voltage lighting with transformers

• dimmer switches

Removing these should eliminate 80%-90% of dirty electricity. To get rid of the remaining requires removing or turning off devices that have low-voltage transformers. Examples are appliances that contain digital clocks and displays and ones with internal computer circuits that are always on. The solution is to turn the power off to these when they are not being used. Either have an electrician install light switches to shut the power off to these when they are not being used or install temporary switches for the following:

• the dishwasher

• a microwave oven

• cordless phones

• flat panel TVs

We are almost there. Cell phones, computers, and similar items cause dirty electricity. The remedy is to use a charging station with a switch to shut the power off at night. Use a power strip and put it on a timer.

We're down to the last 15% of dirty electricity. To get rid of the rest requires shielding the cords for lamps, as discussed in the section on reducing electric fields.

If you do all this, you will have eliminated 98% of dirty electricity. The remaining 2% will come from neighbors connected to the same transformer. Fields decrease with wire distance. The further you are from your neighbors, the better.

There are instances when the electric company is to blame. If, with the power off, you are sensitive to what you believe is electrical noise, have the power tested for excess noise coming from the electric company. Standard (EEE 519) prohibits over 5% total noise on the 120 V feed. You may object and request that the electric company lower it. Some people report being sensitive when it's lower than 3%. A remedy is to install filters at the main electric panel for the house. Ask an electrician to install an outlet at the first breaker on each busbar and to plug a filter into each of the outlets.

Filters

Filters are available to reduce dirty electricity. They are supposed to be plugged into at least one outlet served by each circuit breaker. They do not remove dirty electricity. Capacitors inside them absorb, store, and release the current in a smooth manner. Some people report miracles when they are installed; others report frustration because they don't feel a difference. The reason some may not notice a difference is that the sources of dirty electricity were not removed. Instead of removing the sources, a homeowner may just plug in filters.

Filters create magnetic fields close to them. Do not put one behind or next to a bed unless a gauss meter is used to measure the field to ensure that it is far enough away.

Getting Rid of Fields

Getting Rid of High Frequency Fields

Fields from high frequency sources are often bigger inside a house than outside. Consider that your phone is transmitting its location and updating apps when you're not making calls. Few turn their phones off. Computers are similar. A computer might be in sleep mode, still on. A wireless keyboard or mouse is constantly broadcasting, ready to spring into action should any key be touched.

Get a router without WiFi

Ask your internet service provider for a router that does not have WiFi. It's the only way to make sure there are no fields. A WiFi router will transmit even if the WiFi button is off. Get two routers—one with WiFi, to be turned on when there are guests, and the other without WiFi capability. Use an ethernet cable to connect the router without WiFi capability to your computers.

Case study

A survey was done in a residential home to measure EMFs. The three biggest sources are listed below. WiFi was the strongest source, even though the client was not using WiFi. They had an ethernet cable plugged into their computer, and the WiFi button on the router was turned off. Even though the WiFi was switched off, the router was emitting power on the WiFi frequency. The router needed to be unplugged to get rid of the field.

Average Power Density Measured in Home (nW/cm²)			
	Frequency	WiFi switch on router was turned off	Router was unplugged
WiFi	2.4 - 2.6 GHz	13	0.35
FM Radio	88 - 108 MHz	1.6	1.4
Cellular	824 - 849 MHz	0.12	0.10

When the router was unplugged, the WiFi fell to third place, with 0.35 nW/cm² of power. The residual was coming from the neighbor's WiFi. The neighbor's contribution was 3% compared to the 97% from the router inside the house that had the WiFi switched off. The next biggest source was FM radio, followed by cell phones and the cellular network.

Some cities have precautionary limits based on the peak, maximum pulse observed instead of the average exposure. Here are the measurements for the peaks.

High Frequency Fields

WiFi was still the top source when the router was plugged in, even though the WiFi switched was turned off. The WiFi router had a higher peak than the cellular network. When the router was unplugged, cell phones and the cellular network became the highest sources. The pulsed (peak, maximum) readings from cellular sources were not as high as the WiFi. The pulsed emission from the WiFi was 300 times greater than those from cellular sources.

Peak (Maximum) Density in Home (nW/cm²)			
	Frequency	WiFi switch on router was turned off	Router was unplugged
WiFi	2.4 - 2.6 GHz	380	0.37
FM Radio	88 - 108 MHz	14	2.2
Cellular	824 - 849 MHz	2.6	11

Another house had two routers, one for high speed (5 GHz) and the other for traditional speed (2.4 GHz) internet. Some people worry that high speed internet creates more radiation since the frequency is higher. Although the average power is higher for high-speed WiFi, the traditional speed internet router had a higher peak and maximum emission, making it a greater hazard.

Here are some ways to lower exposure to high-frequency radiation:

• Get a router without WiFi. Connect the router

directly to your computer using an ethernet cable.

- Avoid or limit the use of bluetooth devices.

- For time machine backups, use a hard drive that does not have WiFi capability, and connect it directly to the computer.

- Turn off the keyboard and mouse when a computer is not being used.

- Get a corded phone.

Getting Rid of Artificial DC Magnetic Fields

Get a bed without metal

Metal springs and bed frames distort the Earth's magnetic field. Springs are troublesome, as they change the field in random, haphazard directions. Every spring creates its own, localized change in the direction of the field.

The simple way to assess a mattress with metal springs is with a compass, the kind used for hiking. A phone will not work because the compass in a phone uses a GPS system, not Earth's magnetic field. Pull the bed sheets off. Place the compass flat on the mattress. Slowly move the compass across the mattress. You will see the needle move.

Don't put solar panels near bedrooms

Solar power generates a DC current. The inverter that converts DC to AC electricity that can be used in the home creates magnetic fields and dirty electricity. If you

153

want to reduce EMFs, the safest thing is not to have solar.

Shielding

Shielding

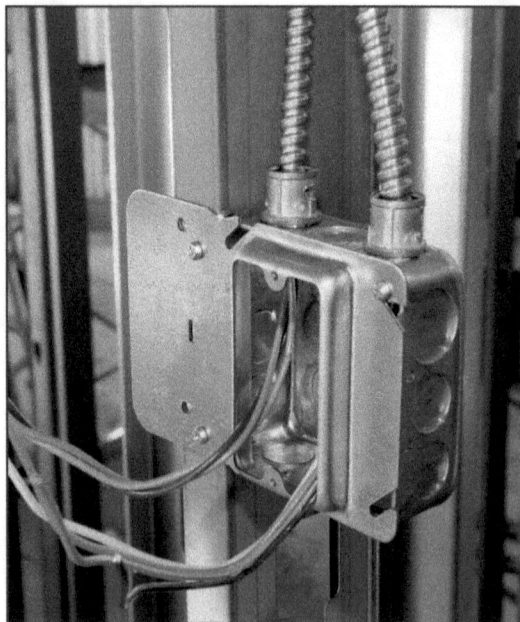

Shielding Electric Fields

Metal conduit

The effective way to reduce or practically eliminate electric fields is to install the wires inside metal conduit or use a metal-jacketed type of wire. Since the metal is grounded, the wires inside the conduit are shielded. Use Health Care Facility (HCF) metal clad armored cabling or Electrical Metallic Tubing (EMT). The outer metal casing must serve as an equipment grounding conductor. Electrical boxes and bushings must also be metal. Cables which are 240 V have two 120 V hot conductors that are of opposite polarity, thus canceling the electric field without the need for a special type of cable.

Romex is the standard type of wire used for wiring a residential home. It has a PVC jacket which does not provide shielding. Romex was created in 1922 by the Rome Wire Company and is used as a generic reference to non-metallic types of wire.

An electrician may say it's difficult and expensive to install metal conduit instead of romex. Find an electrician who wires commercial buildings. Commercial

Shielding

buildings are wired using metal conduit.

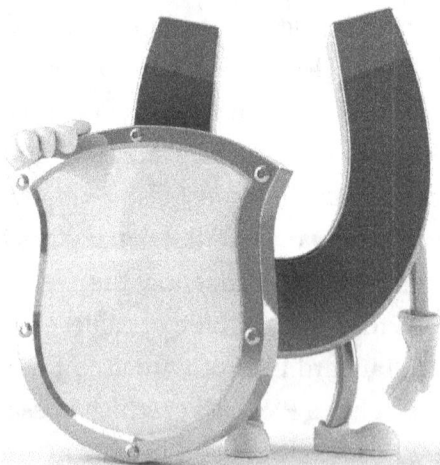

Shielding Magnetic Fields

Mu-metal

Mu-metal is a nickel-based alloy that shields magnetic fields. According to Wikipedia, the name "Mu" came from the Greek letter mu, which represents permeability in physics and engineering. Mu-metal contains approximately 80% nickel, 15% iron, 5% molybdenum, copper, chromium, and other elements.

Shields made with mu-metal do not block magnetic fields. Magnetic fields bend around it, concentrating at the corners. If, for example, you place a flat piece of shielding on a wall to cover a fuse box, the magnetic field will be amplified at the corners. The most effective shape for shielding is a closed container (Faraday cage). For it to be effective, you must put the fuse box inside a metal box made with the shielding material. The applications in a residential environment are therefore limited.

Motor magnet

I had a 1980s Toyota van, and I wanted to shield the mag-

Shielding

netic field created by the alternator on the engine. The seats were on top of the engine, so I took the seats out and installed Mu-metal over the top of the engine as best I could. I put the seats back in, started the engine, and took a reading. The field did not change. There were too many crinkles and corners in the shielding or the field simply went around the corners. I spent a few hundred dollars on something that made no difference.

Magnetic Fields

Shielding

Shielding High Frequency Fields

Imagine you want to sleep in and are sensitive to light. You might install blackout blinds, but it can be difficult to make it completely dark. It's the same with radio waves. High frequency waves behave in the same way as visible light; they find a way to bounce around and get inside. You might be able to reduce the level, but it's almost impossible to eradicate the EMFs completely.

Shielding can make a difference. Windows are the most important location to consider. Focus your efforts there. Expensive shielding materials are not required to make some difference. Standard curtains, blinds, window tints, and metal window screens all block visible light. Since high frequency EMFs adhere to the same laws of physics as light, these block or absorb some amount of man-made radiation.

Be wary of consultants who take measurements and sell products. It's a conflict of interest. Have a consultant take readings and write you a report. Order shielding based on the readings if you decide it's necessary.

Shielding

Carbon-based paint (Y-Shield)

You could paint the house with a special paint that contains nano-sized carbon particles. (Carbon conducts better than copper.) The product comes with tape to embed in the paint to be connected to a grounding rod.

There are a few things to consider. There may be an odor, and the finished wall will not be as vapor-permeable. The manufacturer recommends protecting the coating by painting over it with a weatherproof paint. My general concern with shielding is the potential for other fields to interact or be attracted, causing one field to go down and another to go up. Although I can't foresee problems the paint may create, I have no experience using it.

Blinds and tinted windows

Consult with the manufacturers of these materials to see what frequencies are shielded. Purchase something based on what you are trying to block: radio, TV transmitters on a hill, or a cell phone antenna across the street.

Keep the cage clean

If you are going to install shielding, consider removing every energy-saving light bulb in the house. Inside each one is a diode circuit that emits RFs. The shielding you are installing to block outside sources, if it is effective, will trap sources inside, increasing radiation. EMFs reflect off walls and surfaces just as visible light does.

Brainteaser

The door on a microwave oven has a screen you can look through to see what's inside. Does that mean some radiation is coming through? No. The door screen has a metal piece of shielding with small holes too small for the waves, the size of the frequency the microwave operates at, to pass. If you place a cell phone in the microwave and have someone call it, it will ring. Phones operate at other frequencies. If all the frequencies were blocked, you would not be able to see through the door, as it would block the frequencies of visible light too.

Shielding

Gimmicks

A gimmick is something that does not reduce the fields as measured using a professional meter. I have measured fields with and without such products. Some products that are sold to shield EMFs (those you plug in to power them) emit new frequencies and cause fields to increase.

- Anything you plug in to work is a gimmick, with the exception of ground sheets plugged into a tent stake outside.

- Anything you have to wear to be protected (a piece of jewelry, bracelet, or charm) is a gimmick. This excludes the earlier recommended anti-static wrist strap that workers in the electronics industry wear to prevent static electricity.

- Stickers and objects that "shield" are gimmicks, with the exception of Mu-metal used to shield magnetic fields.

Instead of buying gimmicky products, consider limiting the time you spend near a source if you are unable to turn it off.

What does zero mean?

Saying zero-EMF implies that all EMFs are shielded when it's only certain frequencies. I saw an ad for a product that claimed to be a Zero-EMF shelter, but a phone worked inside the shelter. Magnetic fields cannot be reduced. Unless the shielding is grounded, electric fields from house wiring will not be reduced. Some shielding acts as an antenna and attract fields.

Neutralizers

Be wary of products sold to neutralize EMFs by broadcasting the Earth's natural field. Plugging something in creates new and artificial fields. The exact frequency of the earth cannot be replicated. Do not introduce new fields in an attempt to cancel those you don't care for.

If it sounds too good...

I saw a commercial for a bracelet that is supposed to protect the person wearing it from EMFs. The commercial showed people being muscle tested. If a body does not have a negative reaction to something, that does not mean it is protected from EMFs. The manufacturers of

Gimmicks

these types of products usually base their claims on how people feel, and they may measure blood pressure and heart rate. There is no reduction in fields measured with a meter.

Try this instead

I had a client who bought a protection charm for a spouse for their birthday. Their partner did not like to wear it because he said it made him feel worse. We recruited a chiropractor trained in kinesiology to perform an experiment to test different people wearing the charm. The study was performed by turning the power on and off randomly, without letting people know if the power was on or off. Subjects tested the strongest when they said something such as the following, while not wearing the charm:

I now surround myself in a bubble of protection.
I am protected from electromagnetic fields.
I am surrounded by a bubble of protection.

Supplementary

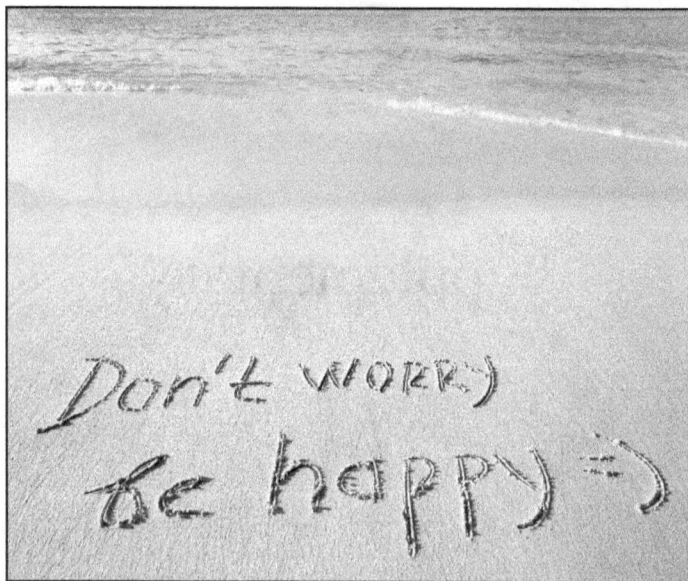

Everyday Tips for Reducing EMFs

Do this for a good night's sleep

Unplug everything next to the bed. Get a battery operated alarm clock. Try moving the bed or sleeping in another room.

Do this when remodeling

Replace romex wiring with metal conduit wire. Install kill switches for appliances to shut them off when they are not being used. Hard-wire rooms with an ethernet cable so the WiFi can be shut off.

Do this in the office

Get a phone with a cord, one without a power connection. Similar to WiFi routers, the field emitted from the base station of a cordless phone is always present, even when not making calls. In a house where there was a field of a few mG at the desk, there were two power strips. Both were a few feet away and not contributing to a high field. The source was the transformer in the cordless phone.

Change the light bulbs

Energy-saving light bulbs are an example of a new technology that created a nightmare for people, pets, wildlife, and the environment. The bulbs contain mercury, which 9 V batteries were once made with. They had a higher capacity than standard carbon-zinc batteries, a more stable voltage, and were used in long-life applications. Mercury was phased out of batteries because it is an environmental pollutant. Consider the mercury contamination that results inside a home when a bulb is broken.

The bulbs flicker at frequencies beyond what a normal person notices, cause headaches, and disturb concentration. Some can see the flicker. Pets hear it. Although LEDs are better than CFLs, as they do not contain mercury, they also flicker and produce fields. The electronic circuit is the culprit.

Common Sources of EMFs— Myth and Fact

Computers

Computers, especially laptops, are not big sources. The power supply may be a source. A grounded metal chassis, such as those used on laptops made by Apple, reduces the electric field.

New cars

The batteries and electronics in hybrid cars might be responsible for large magnetic fields. The alternators in all cars produces a magnetic field. The field dissipates within a few feet. Bigger cars have lower readings because the alternator is further from the seats, towards the front of the car. People in small cars may have larger exposures. Use a gauss meter to measure it.

Microwave ovens

Microwave ovens produce radiation (high frequency fields) to heat food. The highest radiation is found at the door from leakage. You can buy a cheap microwave oven leak detector to check, but there's no need. Almost all ovens leak at the door seal, so do not stand near it when cooking.

When a microwave is not cooking, there is no hazard. The magnetic field detected is from the LCD display, the same type of EMF from a clock radio, and dissipates within a few feet.

Fuse box

The fields at fuse boxes usually dissipate within an arm's length distance. Use the three-foot rule; avoid putting a bed or couch within three feet.

Power lines

You can't tell by looking at power lines how big the fields are. It's always preferable if they are on the other side of the street. Burying power lines does not shield fields.

EMFs—Myth and Fact

Television sets

Flat screen TVs solved a big problem. TVs and computer monitors used to have cathode ray tubes. It used to be that it was not the person sitting in front of a computer, but the person in the cubicle behind the monitor who was affected because radiation went out the back.

Granite countertops

Some granite countertops emit radiation. The strength dissipates quickly. Since the affected granite is the size of a quarter, you'd have to get lucky to find the spot and sleep on it for a few years to have significant exposure. If you find such a spot, you could have the granite company cut it out. Why bother?

In general, radiation from natural sources increases with altitude from sea level, where it is lowest. It consists of alpha particles, beta particles, and gamma radiation. Alpha particles are made of protons and neutrons—bowling ball sized particles that are easily blocked by a piece of paper. Beta particles are high-speed electrons, like a bullet. They may be blocked, but they require a thicker material. Gamma radiation is electromagnetic energy that thick concrete walls are required to shield.

Most Geiger counters only give a total reading of the alpha, beta, and gamma emissions. To read gamma radiation, you need a Geiger counter as worn by workers at a nuclear power plants.

Glossary

Electric field: Electric fields are present if the power is on, even if nothing is turned on. The common source is house wiring. Measured using body voltage.

Magnetic field: Created only when current flows. If the wires are properly installed, current will be balanced, and there will be no net magnetic field in a house due to wires. Measured using a Gauss meter.

Volts (V): The measure of energy in electricity.

Current (I): The rate that electricity is flowing. Measured in amps.

Neutral: The white wire. Current returns to the fuse box on it.

Ground: The bare copper wire. It should never have current on it.

Glossary

Radiation: Light radiates light. Heat radiates heat. EMFs radiate electric and magnetic fields.

Bau-biology: An organization founded in Germany that trains people how to test homes. After being introduced in the United States, the American organization re-named itself "Building Biology."

Further Reading

EMF Handbook. Understanding and Controlling Electromagnetic Fields in Your Life. Stephen Prate. Wait Group Press, 1993. This is a classic. Prata also wrote the first book I read on C++ programming when I was a young engineer.

Tracing EMFs in Building Wiring and Grounding. Karl Riley. Magnetic Sciences International, 1995. One of the first texts written about how to identify sources of magnetic fields and fix the wiring errors that create them. The focus is on magnetic fields.

Electrical Systems and their Fields of (Biological) Influence. Sal La Duca, 2006. Contains technical details on how power is generated and distributed, how fields are created, and how to fix problems.

References

Questions and Answers About EMF, Electric and Magnetic Fields Associated with the Use of Electric Power. National Institute of Environmental Health Services and U.S. Department of Energy. U.S. Government Printing Office, January, 1995.

"School loses magnetic hot spot." Magruder, Janie. *The Mesa Tribune*, January, 1991.

"A Brief Chronology of 'Stray Voltage' Developments on Dairy Farms in Wisconsin. The Perspective of a Dairy Farmer and an Electrical Engineer." Spark Burmaster. October, 1998.

"Importance of Addressing National Electric Code Violations That Result in Unusual Exposure to 60 HZ Magnetic Fields." Jack Adams, J. Samuel Bitler and Karl Riley. *Bioelectromagnetics* 25:102-106, 2004.

References

"Cows are rebounding following transformer relocation." *The County Today*. January 17, 1996.

"Electrical grounding, pipe integrity, and shock hazard, Executive Summary." S.J. Duranceau et al. Journal AWWA. Volume 90, Issue 7, July, 1998.

"Understanding Ground Currents: An Important Factor in Electromagnetic Exposure." Duane A. Dahlberg. The Electromagnetic Research Foundation. January, 2000.